ACTIVE PHYSICS® | ACTIVE CHEMISTRY™ | ACTIVE BIOLOGY™ | EARTHCOMM®

UNIT 3 ACTIVE BIOLOGY ▪ TEACHER'S EDITION

Coordinated Science for the 21st Century™

An Integrated, Project-Based Approach

Arthur Eisenkraft, Ph.D.
Ruta Demery
Gary Freebury
Robert Ritter, Ph.D.
Michael Smith, Ph.D.
John B. Southard, Ph.D.

84 Business Park Drive, Armonk, NY 10504 Phone (914) 273-2233
Fax (914) 273-2227 Toll Free (888) 698-TIME (8463) www.its-about-time.com

It's About Time
President
Tom Laster

Director of Product Development
Barbara Zahm, Ph.D.

Creative/Art Director
John Nordland

Design/Production
Kathleen Bowen
Burmar Technical Corporation
Nancy Delmerico
Kadi Sarv
Jon Voss

Illustrations
Tomas Bunk
Dennis Falcon

Project Editor
Ruta Demery
EarthComm, Active Physics, Active Chemistry, Active Biology, Coordinated Science for the ***21st Century***

Project Managers
Ruta Demery
EarthComm, Active Physics
Barbara Zahm
Active Physics
Active Chemistry, Active Biology, Coordinated Science for the ***21st Century***

Project Coordinators
Loretta Steeves
Coordinated Science for the ***21st Century***
Emily Crum
Matthew Smith
EarthComm

Technical Art
Stuart Armstrong
EarthComm
Burmar Technical Corporation
Kadi Sarv
Active Physics, Active Chemistry
Active Biology

Photo Research
Caitlin Callahan
Kathleen Bowen
Jon Voss
Kadi Sarv
Jennifer Von Holstein

Safety Reviewers
Ed Robeck, Ph.D.
EarthComm, Active Biology
Gregory Puskar
Active Physics
Jack Breazale
Active Chemistry

Printed and bound in the United States of America

ISBN #1-58591-357-X 5 Volume Set ISBN #1-58591-354-5

1 2 3 4 5 VH 09 08 07 06 05

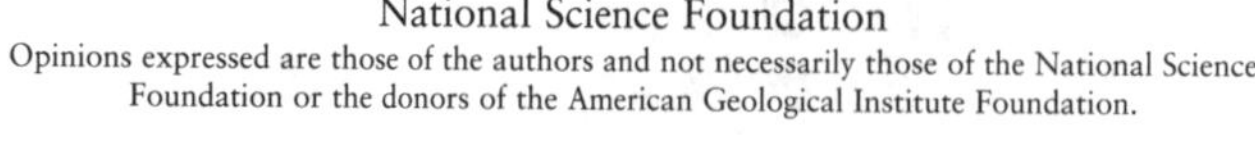

This project was supported, in part, by the
National Science Foundation
Opinions expressed are those of the authors and not necessarily those of the National Science Foundation or the donors of the American Geological Institute Foundation.

UNIT 3 ACTIVE BIOLOGY ▪ TEACHER'S EDITION

Coordinated Science for the 21st Century™

An Integrated, Project-Based Approach

Coordinated Science for the **21st Century** is an innovative core curricula assembled from four proven inquiry-based programs. It is supported by the National Science Foundation and was developed by leading educators and scientists. Unit 1, *Active Physics,* was developed by the American Association of Physics Teachers and the American Institute of Physics.

Both *Active Physics* and *Active Chemistry,* Units 1 and 2, are projects directed by Arthur Eisenkraft, Ph.D., past president of the NSTA.

Active Biology, Unit 3, was developed to follow the same Active Learning Instructional Model as *Active Physics* and *Active Chemistry.*

EarthComm, Unit 4, was developed by the American Geological Institute, under the guidance of Michael Smith, Ph.D., former Director of Education and Outreach, and John Southard, Ph.D., of MIT.

Each unit of this course has been designed and built on the National Science Education Standards. Each utilizes the same instructional model and the same inquiry-based approach.

Project Director, Active Physics and Active Chemistry

Arthur Eisenkraft has taught high school physics for over 28 years and is currently the Distinguished Professor of Science Education and a Senior Research Fellow at the University of Massachusetts, Boston. Dr. Eisenkraft is the author of numerous science and educational publications. He holds U.S. Patent #4447141 for a Laser Vision Testing System (which tests visual acuity for spatial frequency).

Dr. Eisenkraft has been recognized with numerous awards including: Presidential Award for Excellence in Science Teaching, 1986 from President Reagan; American Association of Physics Teachers (AAPT) Excellence in Pre-College Teaching Award, 1999; AAPT Distinguished Service Citation for "excellent contributions to the teaching of physics", 1989; Science Teacher of the Year, Disney American Teacher Awards in their American Teacher Awards program, 1991; Honorary Doctor of Science degree from Rensselaer Polytechnic Institute, 1993; Tandy Technology Scholar Award 2000.

In 1999 Dr. Eisenkraft was elected to a 3-year cycle as the President-Elect, President and Retiring President of the National Science Teachers Association (NSTA), the largest science teacher organization in the world. In 2003, he was elected a fellow of the American Association for the Advancement of Science (AAAS).

Dr. Eisenkraft has been involved with a number of projects and chaired many competition programs, including: the Toshiba/NSTA ExploraVisions Awards (1991 to the present); the Toyota TAPESTRY Grants (1990 to the present); the Duracell/NSTA Scholarship Competitions (1984 to 2000). He was a columnist and on the Advisory Board of *Quantum* (a science and math student magazine that was published by NSTA as a joint venture between the United States and Russia; 1989 to 2001). In 1993, he served as Executive Director for the XXIV International Physics Olympiad after being Academic Director for the United States Team for six years. He has served on a number of committees of the National Academy of Sciences including the content committee that helped write the National Science Education Standards.

Dr. Eisenkraft has appeared on *The Today Show*, *National Public Radio*, *Public Television*, *The Disney Channel* and numerous radio shows. He serves as an advisor to the ESPN Sports Figures Video Productions.

He is a frequent presenter and keynote speaker at National Conventions. He has published over 100 articles and presented over 200 papers and workshops. He has been featured in articles in *The New York Times*, *Education Week*, *Physics Today*, *Scientific American*, *The American Journal of Physics* and *The Physics Teacher*.

Content Specialist, Active Chemistry

Gary Freebury, a noted chemistry teacher, educator, and writer worked as the Project Manager and Editor for the 3 Prototype chapters of *Active Chemistry* which are currently being field tested but are also in print, and he will continue to serve as the Project Manager and Editor to the project. He will be responsible for the writing of any introductory materials, producing the table of contents, indices, glossary, and reference materials. He will have a critical role in maintaining the integrity of safety standards across all units. He will also coordinate any modifications, changes, and additions to the materials based upon pilot and field-testing results.

Mr. Freebury has been teaching chemistry for more than 35 years. He has been the Safety Advisor for Montana Schools, past director of the Chemistry Olympiad, past chairman of the Montana Section of the American Chemical Society (ACS), member of the Executive Committee of the Montana Section of the ACS, and a past member of the Montana Science Advisory Council. Mr. Freebury has been the regional director and author of Scope, Sequence and Coordination (SS&C) – Integrated Science Curriculum and Co-director of the National Science Foundation supported Chemistry Concepts four-year program. He earned a B.S. degree at Eastern Montana College in mathematics and physical science, and an M.S. degree in chemistry at the University of Northern Iowa.

Principal Investigator, EarthComm

Michael Smith, Ph.D., is a former Director of Education at the American Geological Institute in Alexandria, Virginia. Dr. Smith worked as an exploration geologist and hydrogeologist. He began his Earth Science teaching career with Shady Side Academy in Pittsburgh, PA in 1988 and most recently taught Earth Science at the Charter School of Wilmington, DE. He earned a doctorate from the University of Pittsburgh's Cognitive Studies in Education Program and joined the faculty of the University of Delaware School of Education in 1995. Dr. Smith received the Outstanding Earth Science Teacher Award for Pennsylvania from the National Association of Geoscience Teachers in 1991, served as Secretary of the National Earth Science Teachers Association, and is a reviewer for Science Education and The Journal of Research in Science Teaching. He worked on the Delaware Teacher Standards, Delaware Science Assessment, National Board of Teacher Certification, and AAAS Project 2061 Curriculum Evaluation programs.

Senior Writer, EarthComm

John B. Southard, Ph.D., received his undergraduate degree from the Massachusetts Institute of Technology in 1960 and his doctorate in geology from Harvard University in 1966. After a National Science Foundation postdoctoral fellowship at the California Institute of Technology, he joined the faculty at the Massachusetts Institute of Technology, where he is currently Professor of Geology Emeritus. He was awarded the MIT School of Science teaching prize in 1989 and was one of the first cohorts of the MacVicar Fellows at MIT, in recognition of excellence in undergraduate teaching. He has taught numerous undergraduate courses in introductory geology, sedimentary geology, field geology, and environmental Earth Science both at MIT and in Harvard's adult education program. He was editor of the Journal of Sedimentary Petrology from 1992 to 1996, and he continues to do technical editing of scientific books and papers for SEPM, a professional society for sedimentary geology. Dr. Southard received the 2001 Neil Miner Award from the National Association of Geoscience Teachers.

Primary Author, Active Biology
Project Editor, EarthComm, Active Physics, Active Chemistry and Active Biology

Ruta Demery has helped bring to publication several National Science Foundation (NSF) projects. She was the project editor for *EarthComm*, *Active Physics*, *Active Chemistry*, and *Active Biology*. She was also a contributing writer for *Active Physics* and *Active Biology*, both students' and teachers' editions. Besides participating in the development and publishing of numerous innovative mathematics and science books for over 30 years, she has worked as a classroom science and mathematics teacher in both middle school and high school. She brings to her work a strong background in curriculum development and a keen interest in student assessment. When time permits, she also leads workshops to familiarize teachers with inquiry-based methods.

Contributing Author, Active Biology and Active Physics

Bob Ritter is presently the principal of Holy Trinity High School in Edmonton, Alberta. Dr. Ritter began his teaching career in 1973, and since then he has had a variety of teaching assignments. He has worked as a classroom teacher, Science Consultant, and Department Head. He has also taught Biological Science to student teachers at the University of Alberta. He is presently involved with steering committees for "At Risk High School Students" and "High School Science." Dr. Ritter is frequently a presenter and speaker at national and regional conventions across Canada and the United States. He has initiated many creative projects, including establishing a science-mentor program in which students would have an opportunity to work with professional biologists. In 1993 Dr. Ritter received the Prime Minister's Award for Science and Technology Teaching. He has also been honored as Teacher of the Year and with an Award of Merit for contribution to science education.

UNIT 3: ACTIVE BIOLOGY

Primary and Contributing Authors

Ruta Demery

Bob Ritter, Ph.D.

Reviewers

Arthur Eisenkraft, Ph.D.
Project Director of *Active Physics* and *Active Chemistry*
Past President of the National Science Teachers Association (NSTA)

William H. Leonard Ph.D.
Clemson University
Professor of Education and Biology
Co-Author *Biology: A Community Context*

Philip Estrada
Biology Teacher
Hollywood High School, LAUSD

Marissa Hipol
Biology Teacher
Hollywood High School, LAUSD

Laura Hajdukiewicz
Biology Teacher
Andover, MA

Your students can do science. Here are the reasons why . . .

When science learning is based on functional use and active involvement, students get turned on to science. The Active Learning Instructional Model outlined below gives **all** students the opportunity to succeed. Whether it's physics, chemistry, biology, or Earth Science, your students learn science, as they see how science works for them every day and everywhere.

Look for these features in each chapter of *Coordinated Science for the **21st Century***.

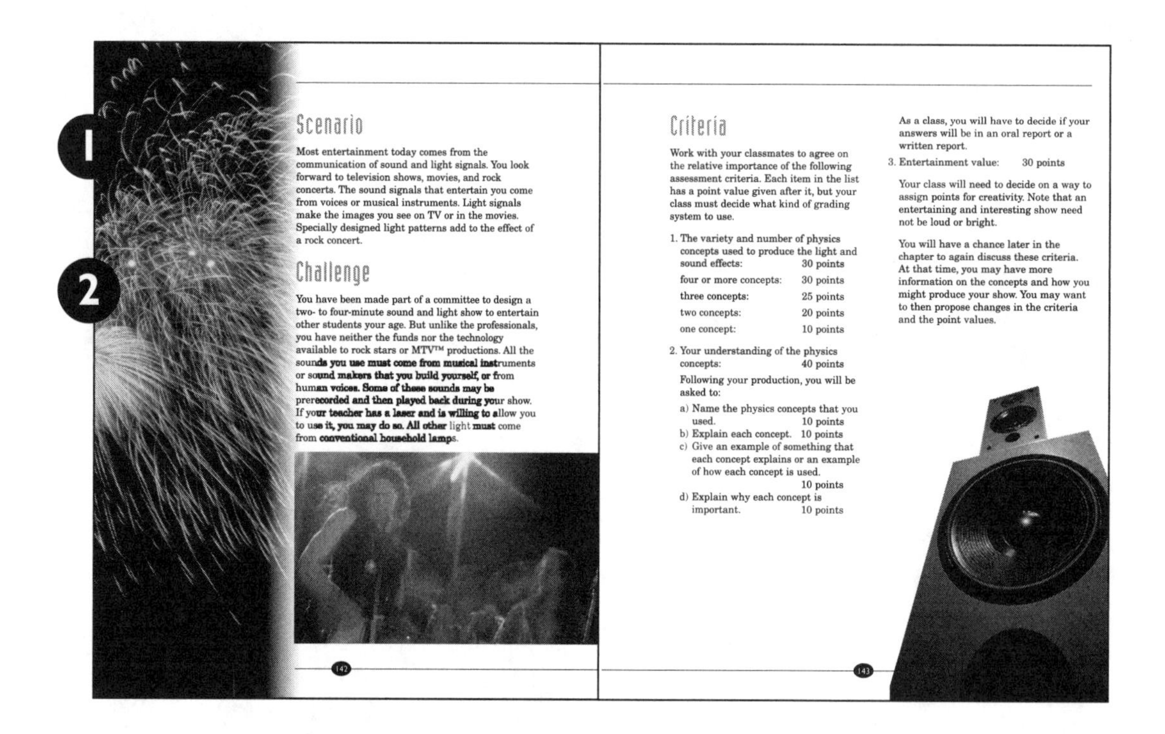

1

2

Scenario

Most entertainment today comes from the communication of sound and light signals. You look forward to television shows, movies, and rock concerts. The sound signals that entertain you come from voices or musical instruments. Light signals make the images you see on TV or in the movies. Specially designed light patterns add to the effect of a rock concert.

Challenge

You have been made part of a committee to design a two- to four-minute sound and light show to entertain other students your age. But unlike the professionals, you have neither the funds nor the technology available to rock stars or MTV™ productions. All the sounds you use must come from musical instruments or sound makers that you build yourself, or from human voices. Some of these sounds may be prerecorded and then played back during your show. If your teacher has a laser and is willing to allow you to use it, you may do so. All other light must come from conventional household lamps.

Criteria

Work with your classmates to agree on the relative importance of the following assessment criteria. Each item in the list has a point value given after it, but your class must decide what kind of grading system to use.

1. The variety and number of physics concepts used to produce the light and sound effects: 30 points
   - four or more concepts: 30 points
   - three concepts: 25 points
   - two concepts: 20 points
   - one concept: 10 points
2. Your understanding of the physics concepts: 40 points

   Following your production, you will be asked to:

   a) Name the physics concepts that you used. 10 points
   b) Explain each concept. 10 points
   c) Give an example of something that each concept explains or an example of how each concept is used. 10 points
   d) Explain why each concept is important. 10 points

   As a class, you will have to decide if your answers will be in an oral report or a written report.
3. Entertainment value: 30 points

   Your class will need to decide on a way to assign points for creativity. Note that an entertaining and interesting show need not be loud or bright.

You will have a chance later in the chapter to again discuss these criteria. At that time, you may have more information on the concepts and how you might produce your show. You may want to then propose changes in the criteria and the point values.

142

143

1 Scenario

Each chapter begins with a realistic event or situation. Students might actually have experienced the event or can imagine themselves participating in a similar situation at home, in school, or in your community. Chances are your students probably never thought about the science involved in each case, but now they will!

2 Challenge

This feature presents students with a challenge that they are expected to complete by the end of the chapter. This challenge gives them the opportunity to learn science as they produce a realistic, science-based project. As they progress through the chapter they will accumulate all the scientific knowledge they need to successfully complete the challenge.

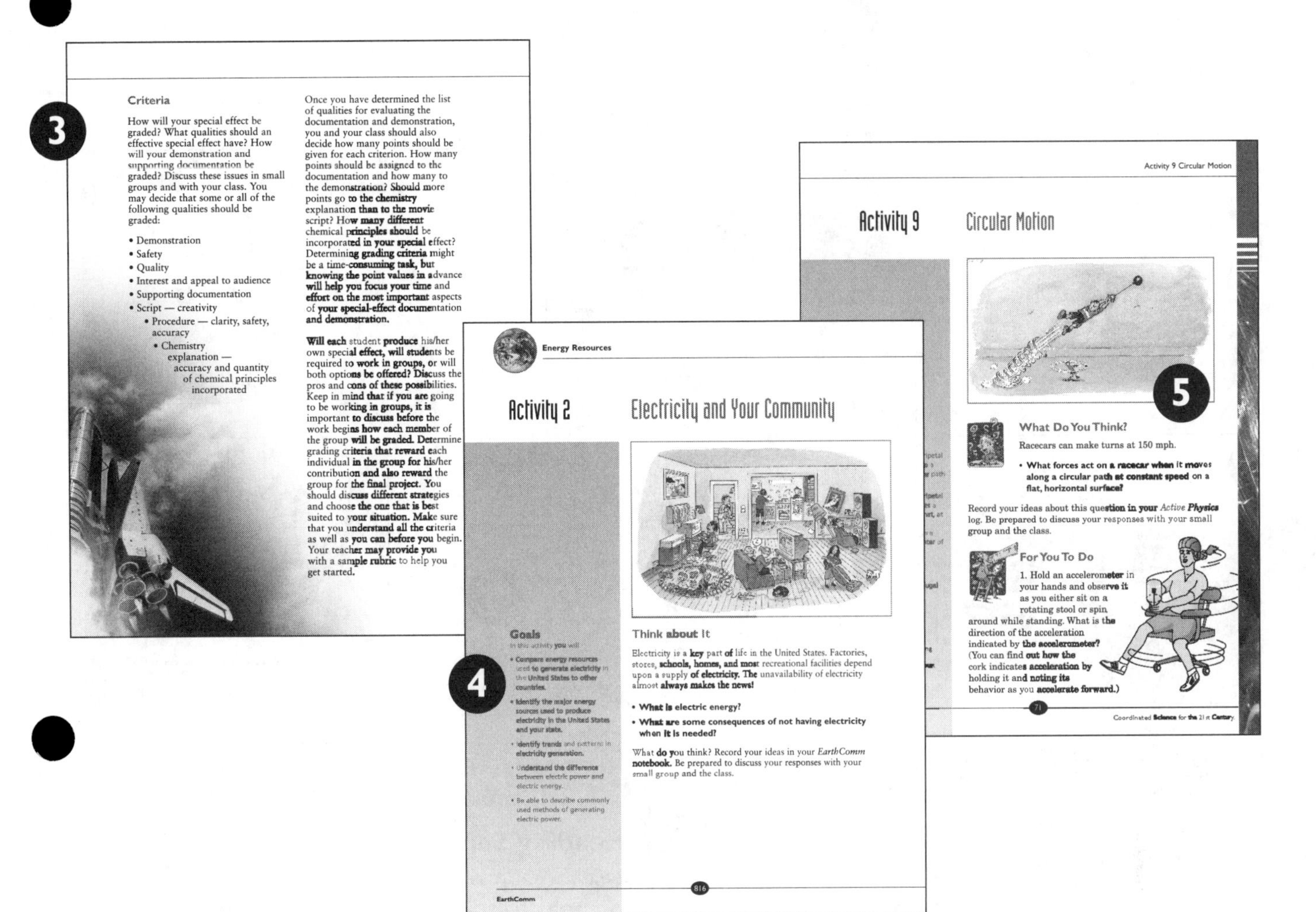

Criteria

How will your special effect be graded? What qualities should an effective special effect have? How will your demonstration and supporting documentation be graded? Discuss these issues in small groups and with your class. You may decide that some or all of the following qualities should be graded:

- Demonstration
- Safety
- Quality
- Interest and appeal to audience
- Supporting documentation
- Script — creativity
  - Procedure — clarity, safety, accuracy
    - Chemistry explanation — accuracy and quantity of chemical principles incorporated

Once you have determined the list of qualities for evaluating the documentation and demonstration, you and your class should also decide how many points should be given for each criterion. How many points should be assigned to the documentation and how many to the demonstration? Should more points go to the chemistry explanation than to the movie script? How many different chemical principles should be incorporated in your special effect? Determining grading criteria might be a time-consuming task, but knowing the point values in advance will help you focus your time and effort on the most important aspects of your special-effect documentation and demonstration.

Will each student produce his/her own special effect, will students be required to work in groups, or will both options be offered? Discuss the pros and cons of these possibilities. Keep in mind that if you are going to be working in groups, it is important to discuss before the work begins how each member of the group will be graded. Determine grading criteria that reward each individual in the group for his/her contribution and also reward the group for the final project. You should discuss different strategies and choose the one that is best suited to your situation. Make sure that you understand all the criteria as well as you can before you begin. Your teacher may provide you with a sample rubric to help you get started.

Energy Resources

Activity 2 Electricity and Your Community

Goals

In this activity you will:

- Compare energy resources used to generate electricity in the United States to other countries.
- Identify the major energy sources used to produce electricity in the United States and your state.
- Identify trends and patterns in electricity generation.
- Understand the difference between electric power and electric energy.
- Be able to describe commonly used methods of generating electric power.

Think about It

Electricity is a key part of life in the United States. Factories, stores, schools, homes, and most recreational facilities depend upon a supply of electricity. The unavailability of electricity almost always makes the news!

- What is electric energy?
- What are some consequences of not having electricity when it is needed?

What do you think? Record your ideas in your *EarthComm* notebook. Be prepared to discuss your responses with your small group and the class.

816

EarthComm

Activity 9 Circular Motion

Activity 9 Circular Motion

What Do You Think?

Racecars can make turns at 150 mph.

- What forces act on a racecar when it moves along a circular path at constant speed on a flat, horizontal surface?

Record your ideas about this question in your *Active Physics* log. Be prepared to discuss your responses with your small group and the class.

For You To Do

1. Hold an accelerometer in your hands and observe it as you either sit on a rotating stool or spin around while standing. What is the direction of the acceleration indicated by the accelerometer? (You can find out how the cork indicates acceleration by holding it and noting its behavior as you accelerate forward.)

71

Coordinated Science for the 21st Century

3 Criteria

Before your students begin the chapter and the challenge, they will explore together, with you and among themselves, exactly how they will be graded. Students thus become involved in evaluating their own learning process.

4 Goals

A list of goals is provided at the beginning of each activity to let students know the learning outcomes expected from their active scientific inquiries.

5 What Do You Already Know?

Before starting each activity students' prior knowledge is tapped and shared as they discuss the introductory questions. Students are specifically told not to expect to come up with the "right" answers, but to share current understandings.

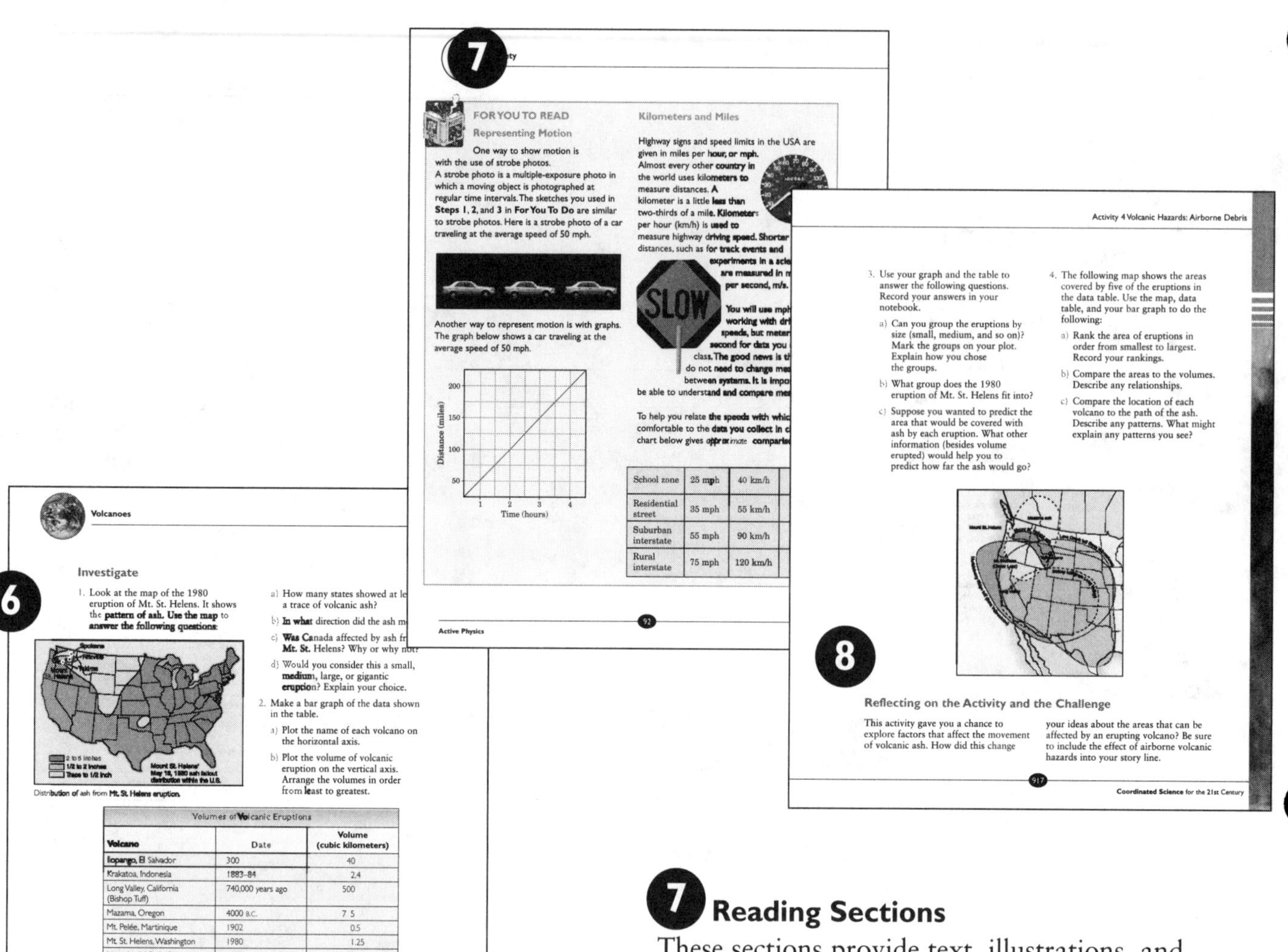

Volumes of Volcanic Eruptions		
Volcano	Date	**Volume (cubic kilometers)**
Ilopango, El Salvador	300	40
Krakatoa, Indonesia	1883–84	2.4
Long Valley, California (Bishop Tuff)	740,000 years ago	500
Mazama, Oregon	4000 B.C.	7 5
Mt. Pelée, Martinique	1902	0.5
Mt. St. Helens, Washington	1980	1.25
Nevado del Ruiz, Colombia	1985	0.025
Pinatubo, Philippines	1991	10
Santorini, Greece	1450 B.C.	6 0
Tambora, Indonesia	1815	150
Valles, New Mexico	1.4 million years ago	300
Vesuvius, Italy	79	3
Yellowstone, Wyoming (Lava Creek Ash)	600,000 years ago	1000

7 Reading Sections

These sections provide text, illustrations, and photographs to solidify the insights students gained in the activity. Equations and formulas are provided with easy-to-understand explanations. Science Words that may be new or unfamiliar are defined and explained. In some chapters, the Checking Up questions are included to guide the reading.

6 Investigate

In *Coordinated Science for the 21st Century*, students learn science by **doing** science. In small groups, or as a class, they will take part in scientific inquiry by doing hands-on experiments, participating in fieldwork, or searching for answers using the Internet or other reference materials.

8 Reflecting on the Activity and the Challenge

Each activity develops specific skills or concepts necessary for the challenge. This feature helps students see this big picture for themselves. It provides a brief summary of the activity and clarifies the purpose of the activity with respect to the challenge. Students thus see each piece of the chapter jigsaw puzzle.

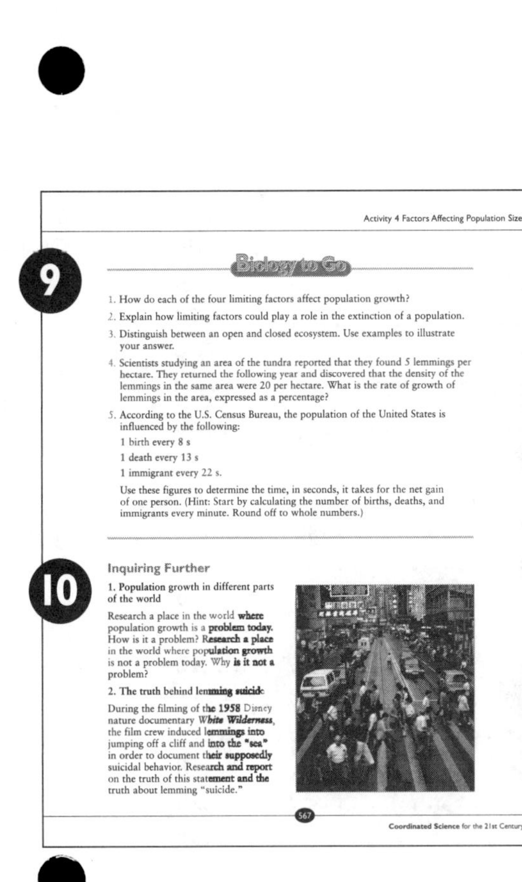
Activity 4 Factors Affecting Population Size

Biology to Go

1. How do each of the four limiting factors affect population growth?
2. Explain how limiting factors could play a role in the extinction of a population.
3. Distinguish between an open and closed ecosystem. Use examples to illustrate your answer.
4. Scientists studying an area of the tundra reported that they found 5 lemmings per hectare. They returned the following year and discovered that the density of the lemmings in the same area were 20 per hectare. What is the rate of growth of lemmings in the area, expressed as a percentage?
5. According to the U.S. Census Bureau, the population of the United States is influenced by the following:
   1 birth every 8 s
   1 death every 13 s
   1 immigrant every 22 s.
   Use these figures to determine the time, in seconds, it takes for the net gain of one person. (Hint: Start by calculating the number of births, deaths, and immigrants every minute. Round off to whole numbers.)

Inquiring Further

1. Population growth in different parts of the world

Research a place in the world where population growth is a problem today. How is it a problem? Research a place in the world where population growth is not a problem today. Why is it not a problem?

2. The truth behind lemming suicide

During the filming of the 1958 Disney nature documentary *White Wilderness*, the film crew induced lemmings into jumping off a cliff and into the "sea" in order to document their supposedly suicidal behavior. Research and report on the truth of this statement and the truth about lemming "suicide."

567 Coordinated Science for the 21st Century

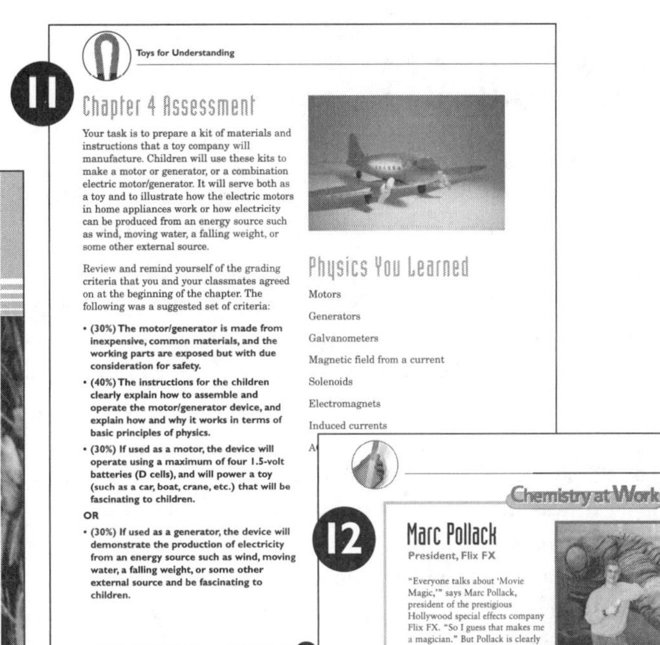
Toys for Understanding

Chapter 4 Assessment

Your task is to prepare a kit of materials and instructions that a toy company will manufacture. Children will use these kits to make a motor or generator, or a combination electric motor/generator. It will serve both as a toy and to illustrate how the electric motors in home appliances work or how electricity can be produced from an energy source such as wind, moving water, a falling weight, or some other external source.

Review and remind yourself of the grading criteria that you and your classmates agreed on at the beginning of the chapter. The following was a suggested set of criteria:

- (30%) The motor/generator is made from inexpensive, common materials, and the working parts are exposed but with due consideration for safety.
- (40%) The instructions for the children clearly explain how to assemble and operate the motor/generator device, and explain how and why it works in terms of basic principles of physics.
- (30%) If used as a motor, the device will operate using a maximum of four 1.5-volt batteries (D cells), and will power a toy (such as a car, boat, crane, etc.) that will be fascinating to children.

OR

- (30%) If used as a generator, the device will demonstrate the production of electricity from an energy source such as wind, moving water, a falling weight, or some other external source and be fascinating to children.

Physics You Learned

Motors
Generators
Galvanometers
Magnetic field from a current
Solenoids
Electromagnets
Induced currents

Active Physics 246

Chemistry at Work

Marc Pollack
President, Flix FX

"Everyone talks about 'Movie Magic,'" says Marc Pollack, president of the prestigious Hollywood special effects company Flix FX. "So I guess that makes me a magician." But Pollack is clearly more comedian than magician. The 'magic' he creates for movies like *Blackhawk Down, Men In Black* and *Cast Away*, in addition to scores of television commercials, museum installments and Las Vegas casinos, is the product not of mysterious hocus-pocus but rather fundamental principles of science. "One of the most important aspects of our work," he continues, "is to push the limits of how chemicals are designed to be used." Among other things, Pollack and his crew at Flix FX use vacuum-forming thermo-plastics to make tin-based silicon molds for everything from prehistoric creatures to futuristic robots. Through a combination of trial and error experimentation and traditional research science, they've perfected the process. "Silicon is what we call an R.T.V.," Pollack explains. "That stands for room temperature vulcanization. So depending on the type and amount of catalyst we use, the mold will cure at different rates and with slightly differing properties." By manipulating the ratio of silicon to catalysts they can make strong, realistic molds in the most efficient way possible. "Increasing the amount of catalyst will speed up the curing process but too much catalyst will shorten the life of the mold," he says. "Every job is different so determining that balance is one of our many challenges."

Pollack, who is now a master in the art of using chemicals like silicon, polypropylene, urethane and urethane elastimers, is not a chemist by trade. He actually graduated from film school at SUNY Purchase in the hopes of becoming "the next Steven Spielberg." Then, through a twist of fate, he became a special effects nut and eventually founded Flix FX in 1990. "Now," Pollack says, "Spielberg may one day come to me."

Special effects — Pollack creates both mechanical and physical — is an industry in a constant state of transformation. "The industry is always trying new stuff and that's exciting," Pollack says. "For instance, someone just developed a great water-based breakaway glass for stunts called Smash Glass. It's similar to fiberglass without the dangerous elements associated with that material and can be made to break into either large chunks or tiny little pieces. I can't wait to get my hands on it and break it over someone's head. That's part of my job these days and I love it."

526 Active Chemistry

9 Science to Go

Questions in this feature ask students to use the key principles and concepts introduced in the activity. Students may also be presented with new situations to apply what they have learned. These questions provide a study guide, helping students review what is most important from the activity. Students will also be given suggestions for ways to organize their work and get ready for the challenge.

10 Inquiring Further

This feature stretches your students' thinking. It provides suggestions for deepening the understanding of the concepts and skills developed in the activity. If you're looking for more challenging or in-depth problems, questions, and exercises, you'll find them right here.

11 Chapter Assessment

How do your students measure up? Here is their opportunity to share what they have actually learned. Using the activities as a guide, they can now complete the challenge they were presented at the beginning of the chapter.

12 Science at Work

Science is an integral part of many fascinating careers. This feature introduces students to people working in fields that involve the principles of science.

Table of Contents

Acknowledgements .iv
Features of *Coordinated Science for the **21st Century*** .viii
The National Science Education Standards (NSES) .xiii
Cooperative Learning .xvi
Reading and Science .xxii
Assessment Opportunities .xxiv
Safety in the Science Classroom .xxvi
The 7E Learning Cycle Model .xxviii

Chapter 9 **A VOTE FOR ECOLOGY** .1
Chapter Overview .2
Chapter Timeline .3
National Science Education Standards .6
Key Science Concepts and Skills .8
Equipment List .10
Scenario and Chapter Challenge .15
Assessment Criteria .17
Assessment Rubric for Chapter Challenge .18
Activity 1: Diversity in Living Things .20
Activity 2: Who Eats Whom? .54
Activity 3: Energy Flow in Ecosystems .86
Activity 4: Factors Affecting Population Size .108
Activity 5: Competition Among Organisms .132
Activity 6: Succession in Communities .152
Activity 7: The Water Cycle .176
Activity 8: Photosynthesis, Respiration, and the Carbon Cycle .206
Activity 9: The Nitrogen and Phosphorous Cycles .230
Alternative End-of-Chapter Assessment .258
Answers to Alternative End-of-Chapter Assessment .260

Chapter 10 **A HIGHWAY THROUGH THE PAST** .263
Chapter Overview .264
Chapter Timeline .265
National Science Education Standards .267
Key Science Concepts and Skills .268
Equipment List .269
Scenario and Chapter Challenge .271
Assessment Criteria .273
Assessment Rubric for Chapter Challenge .274
Activity 1: Adaptations .276
Activity 2: Is It Heredity or the Environment? .302
Activity 3: Natural Selection .322
Activity 4: The Fossil Record .348
Activity 5: Mass Extinction and Fossil Records .374
Alternative End-of-Chapter Assessment .394
Answers to Alternative End-of-Chapter Assessment .396

The National Science Education Standards (NSES)

Coordinated Science for the ***21st Century*** was developed using the Active Learning Instructional Model. This model promotes the style of science instruction that is encouraged by the NSES Standards.

Guide and facilitate learning

- Focus and support inquiries while interacting with students.
- Orchestrate discourse among students about scientific ideas.
- Challenge students to accept and share responsibility for their own learning.
- Recognize and respond to student diversity; encourage all to participate fully in science learning.
- Encourage and model the skills of scientific inquiry as well as the curiosity openness to new ideas and data.
- Promote the healthy scepticism that characterizes science.

Design and manage learning environments that provide students with time, space and resources needed for learning science

- Structure the time available so students are able to engage in extended investigations.
- Create a setting for student work that is flexible and supportive of science inquiry.
- Make available tools, materials, media, and technological resources accessible to students.
- Identify and use resources outside of school.

Engage in ongoing assessment of student learning

- Use multiple methods and systematically gather data about student understanding and ability.
- Analyze assessment data to guide teaching decisions.
- Guide students in self-assessment.

Develop communities of science learners that reflect the intellectual rigor of scientific attitudes and social values conducive to science learning

- Display and demand respect for the diverse ideas, skills, and experiences of all members of the learning community.
- Enable students to have significant voice in decisions about content and context of work and require students to take responsibility for their own learning.
- Nurture collaboration among students.
- Structure and facilitate ongoing formal and informal discussion based on shared understanding of rules.
- Model and emphasize the skills, attitudes and values of scientific inquiry.

Create authentic assessment standards

- Measure those features claimed to be measured.
- Give students adequate opportunity to demonstrate their achievement and understanding.
- Provide assessment that is authentic and developmentally appropriate, set in familiar context, and engaging to students with different interests and experiences.
- Assesses student understanding as well as knowledge.
- Improve classroom practice and plan curricula.
- Develop self-directed learners.

Key NSES Recommendations

*Coordinated Science for the **21st Century*** will bring the key NSES recommendations into your classroom.

Scenario-Driven

There are 15 chapters in *Coordinated Science for the 21st Century*. Each chapter begins with an engaging **Scenario.** This project-based assignment challenges the students and sets the stage for the learning activities and chapter assessments to follow. Chapter contents and activities are selectively aimed at providing the students with the knowledge and skills needed to address this introductory challenge, thus providing a natural content filter in the "less is more" curriculum.

Flexibly Formatted

Chapters are designed to stand alone, so teachers have the flexibility of changing the sequence of presentation of the chapters, omitting an entire chapter, or not finishing all of the chapters.

Multiple Exposure Curriculum

The thematic nature of the course requires students to continually revisit fundamental science principles throughout the year. Students extend and deepen their understanding of these principles as they apply them in new contexts. This repeated exposure fosters the retention and transferability of learning and promotes the development of critical thinking skills.

Constructivist Approach

Students are continually asked to explore how they think about certain situations. As they investigate new situations, they are challenged to either explain observed phenomena using an existing paradigm or to develop a more consistent paradigm. This approach is especially critical in helping students abandon previously held notions in favor of the more powerful ideas and explanations offered by scientists.

Authentic Assessment

For the culmination of each chapter, students are required to demonstrate the usefulness of their newly acquired knowledge by adequately meeting the challenge posed in the chapter introduction. Students are then evaluated on the degree to which they accomplish this performance task. In addition to the project-based assessment, the curriculum also includes other instruments for authentic assessments, as well as non-traditional procedures for evaluating and rewarding desirable behaviors and skills.

Cooperative Grouping Strategies

Use of cooperative groups is integral to the course as students work together in small groups to acquire the knowledge and information needed to address the series of challenges presented through the chapter scenarios. Ample teacher guidance is provided to assure that effective strategies are used in group formation, function, and evaluation.

Math Skills Development/Graphing Calculators and Computer Spreadsheets

The presentation and use of math in *Coordinated Science for the **21st Century*** varies substantially from traditional high school science courses. Math, primarily algebraic expressions, equations, and graphs is approached as a way of representing ideas symbolically. Students begin to recognize the usefulness of math as an aid in exploring and understanding the world around them. Finally, since many of the students in the target audience are insecure about their math backgrounds, the course engages and provides instruction for the use of graphing calculators and computer spreadsheets to provide math assistance.

Minimal Reading Required

Because it is assumed that the target audience reads only what is absolutely necessary, the entire course is activity-driven. Reading passages are presented mainly within the context of the activities and are written at the ninth grade level.

Use of Educational Technologies

Computer software programs extend and enhance the learning opportunities.

- eLabs – Probes and sensors are used to collect data that is then graphed using Data Studio®.
- CPU – Simulated experiments in physical science direct students' inquiry process.
- GETIT – Research activities use catastrophic events to simulate real-life research practices.
- TestGenerator – Crossplatformed software generates tests using a wide variety of assessment criteria.

Problem Solving

For the curriculum to be both meaningful and relevant to the target population, problem-solving related to technological applications and related issues is an essential component of the course. Problem-solving ranges from simple numerical solutions where one result is expected, to more involved decision-making situations where multiple alternatives must be compared.

Challenging Learning Extensions

Throughout the text, a variety of **Stretching Exercises** are provided for more motivated students. These extensions range from more challenging design tasks, to enrichment readings, to intriguing and unusual problems. Many of the extensions take advantage of the frequent opportunities the curriculum provides for oral and written expression of student ideas.

Cooperative Learning

Any classroom using *Coordinated Science for the* ***21st Century*** comes alive with cooperative learning opportunities. The group project-based approach gives students ample opportunities for:

- face-to-face interactions
- respect for diversity
- development of interdependence
- understanding of individual accountability

Benefits of Cooperative Learning

Cooperative learning requires you to structure a lesson so that students work with other students to jointly accomplish a task. Group learning is an essential part of balanced methodology. It should be blended with whole-class instruction and individual study to meet a variety of learning styles and to maintain a high level of student involvement.

Cooperative learning has been thoroughly researched and has been shown to:

- promote trust and risk-taking
- elevate self-esteem
- encourage acceptance of individual differences
- develop social skills
- permit a combination of a wide range of backgrounds and abilities
- provide an inviting atmosphere
- promote a sense of community
- develop group and individual responsibility
- reduce the time on a task
- result in better attendance
- produce a positive effect on student achievement
- develop key employability skills

As with any learning approach, some students will benefit more than others from cooperative learning. Therefore, you may question as to what extent you should use cooperative learning strategies. It is important to involve the student in helping decide which type of learning approaches they prefer, and to what extent each is used in the classroom. When students have a say in their learning, they will accept to a greater extent any method which you choose to use.

Phases of Cooperative Learning Lessons

Organizational Pre-Lesson Decisions

What academic and social objectives will be emphasized? In other words, what content and skills are to be learned and what interaction skills are to be emphasized or practiced?

What will be the group size? Or, what is the most appropriate group size to facilitate the achievement of the academic and social objectives? This will depend on: the amount of individual involvement expected (small groups promote more individual involvement), the task (diverse thinking is promoted by larger groups), the nature of the task or materials available and the time available (shorter time demands smaller groupings to promote involvement).

Who will make up the different groups? Teacher-selected groups usually have the best mix, but this can only happen after the teacher gets to know his/her students well enough to know who works well together. Heterogeneous groupings are most successful in that all can learn through active participation. The duration of the groups' existence may also have some bearing on deciding the membership of groups.

How should the room be arranged? Practicing routines where students move into their groups quickly and quietly is an important aspect. Having students face-to-face is also important. The teacher should still be able to move freely among the groups.

Materials and/or Rewards to be Outlined in Advance

Structure for Positive Interdependence: When students feel they need one another, they are more likely to work together—goal interdependence becomes important. Class interdependence can be promoted by setting class goals, which all teams must achieve in order for class success.

Explanation of the Academic Task: Clear explanation of the academic outcomes is essential. An explanation of the relevance of the activity is also important. Checks for clear understanding can be done either before the groups form or after, but they are necessary for delimiting frustrations.

Explanation of Criteria for Success: Groups should know how their level of success will be determined. Allowing students to play a role in determining the criteria promotes involvement and a sense of vesting.

Specification of Desired Social Behaviors: Definition and explanations of the importance of values of social skills will promote student practice and achievement of these essential life skills.

Structure for Individual Accountability: The use of individual follow-up activities for tasks or social skills promotes individual accountability.

Monitoring and Intervening During Group Work

Through careful monitoring of students' behaviors, intervention can be used most beneficially. Students can be involved in the monitoring by being "a team observer," but only when the students have a very clear understanding of the behavior being monitored.

Interventions can increase the chances for success in completing the activity and can also teach collaborative skills. These interventions should be used as necessary—they should not be interruptions. This means that the facilitating teacher should be moving among the groups as much as possible. During interventions, the problem should be turned back to the students as often as possible, taking care not to frustrate them.

Evaluating the Content and Process of Cooperative Group Work

Assessment of the achievement of content objectives should be completed by both the teacher and the students. Students can go back to their groups after an assignment to review the aspects in which they experienced difficulties.

When assessing the accomplishment of social objectives, two aspects are important: how well things proceeded and where/how improvements might be attempted. Student involvement in this evaluation is a very basic aspect of successful cooperative learning programs.

Organizing Groups

An optimum size of group for most activities appears to be four; however, for some tasks, two may be more efficient. Heterogeneous groups organized by the teacher are usually the most successful. The teacher will need to decide what factors should be considered in forming the heterogeneous groups. Factors which can be considered are: academic achievement, cultural background, language proficiency, sex, age, learning style, and even personality type.

Initially, level of academic achievement may be the simplest way to form groups. Sort the students on the basis of marks on a particular task or on previous year's achievement. Then choose a student from each quartile to form a group. Once formed, groups should be flexible. Continually monitor groups for compatibility and make adjustments as required.

Students should develop an appreciation that it is a privilege to belong to a group. Remove from group work any student who is a poor participant or one who is repeatedly absent. These individuals can then be assigned the same tasks to be completed in the same time line as a group. You may also wish to place a ten percent reduction on all group work that is completed individually.

The chart on the next page presents a variety of group structures and their relative benefits.

Group Structures and Their Functions*

Structure		Brief Description	Academic and Social Functions
Team Building	Round-robin	Each student in turn shares something with his/her teammates.	Expressing ideas and opinions, creating stories. Equal participation, getting acquainted with each other.
Class Building	Corners	Each student moves to a group in a corner or location as determined by the teacher through specified alternatives. Students discuss within groups, then listen to and paraphrase ideas from other groups.	Seeing alternative hypotheses, values, and problem solving approaches. Knowing and respecting differing points of view.
Mastery	Numbered heads together	The teacher asks a question, students consult within their groups to make sure that each member knows the answer. Then one student answers for the group in response to the number called out by the teacher.	Review, checking for knowledge comprehension, analysis, and divergent thinking. Tutoring.
	Color coded co-op cards	Students memorize facts using a flash card game or an adaption. The game is structured so that there is a maximum probability for success at each step, moving from short to long-term memory. Scoring is based on improvement.	Memorizing facts. Helping, praising.
	Pairs check	Students work in pairs within groups of four. Within pairs students alternate — one solves a problem while the other coaches. After every problem or so, the	Practicing skills. Helping, praising.
Concept Development	Three-step interview	pair checks to see if they have the same answer as the other pair.	Sharing personal information such as hypotheses, views on an issue, or conclusions from a unit. Participation, involvement.
	Think-pair-share	Students interview each other in pairs, first one way, then the other. Each student shares information learned during interviews with the group.	Generating and revising hypotheses, inductive and deductive reasoning, and application. Participation and involvement.
	Team word-webbing	Students think to themselves on a topic provided by the teacher; they pair up with another student to discuss it; and then share their thoughts with the class.	Analysis of concept into components, understanding multiple relations among ideas, and differentiating concepts. Role-taking.
Multifunctional	Roundtable	Students write simultaneously on a piece of paper, drawing main concepts, supporting elements, and bridges representing the relation of concepts/ideas.	Assessing print knowledge, practicing skills, recalling information, and creating designs. Team building, participation of all.
	Partners	Each student in turn writes one answer as a paper and a pencil are passed around the group. With simultaneous roundtables, more than one pencil and paper are used.	Mastery and presentation of new material, concept development. Presentation and communication skills.
	Jigsaw	Students work in pairs to create or master content. They consult with partners from other teams. Then they share their products or understandings with the other partner pair in their team.	Acquisition and presentation of new material review and informed debate. Independence, status equalization.

* Adapted from Spencer Kagan (1990), "*The Structural Approach to Cooperative Learning*," Educational Leadership, December 1989/January 1990.

Roles in Cooperative Learning

During a cooperative learning situation, students should be assigned a variety of roles related to the particular task at hand. Following is a list of possible roles that students may be given. It is important that students are given the opportunity of assuming a number of different roles over the course of a semester.

Leader:

Assigns roles for the group. Gets the group started and keeps the group on task.

Organizer:

Helps focus discussion and ensures that all members of the group contribute to the discussion. The organizer ensures that all of the equipment has been gathered and that the group completes all parts of the activity.

Recorder:

Provides written procedures when required, diagrams where appropriate and records data. The recorder must work closely with the organizer to ensure that all group members contribute.

Researcher:

Seeks written and electronic information to support the findings of the group. In addition, where appropriate, the researcher will develop and test prototypes. The researcher will also exchange information gathered among different groups.

Encourager:

Encourages all group members to participate. Values contributions and supports involvement.

Checker:

Checks that the group has answered all the questions and the group members agree upon and understand the answers.

Diverger:

Seeks alternative explanations and approaches. The task of the diverger is to keep the discussion open. "Are other explanations possible?"

Active Listener:

Repeats or paraphrases what has been said by the different members of the group.

Idea Giver:

Contributes ideas, information, and opinions.

Materials Manager:

Collects and distributes all necessary material for the group.

Observer:

Completes checklists for the group.

Questioner:

Seeks information, opinions, explanations, and justifications from other members of the group.

Reader:

Reads any textual material to the group.

Reporter:

Prepares and/or makes a report on behalf of the group.

Summarizer:

Summarizes the work, conclusions, or results of the group so that they can be presented coherently.

Timekeeper:

Keeps the group members focused on the task and keeps time.

Safety Manager:

Responsible for ensuring that safety measures are being followed, and the equipment is clean prior to and at the end of the activity.

Assessment of Cooperative Group Work

Assessment should not end with a group mark. Students and their parents have a right to expect marks to reflect the students' individual contributions to the task. It is impossible for you as the instructor to continuously monitor and record the contribution of each individual student. Therefore, you will need to rely on the students in the group to assign individual marks as merited.

There are a number of ways that this can be accomplished. The group mark can be multiplied by the number of students in the group, and then the total mark can be divided among the students, as shown in the example that follows.

Activity:____________________

Group Mark: 8/10

Number in Group: 4

Total Marks: 32/40

Distribution of Marks

Student's Name	Mark	Signature
Ahmed	8/10	____________________
Jasmin	8/10	____________________
Mike	7/10	____________________
Tabitha	9/10	____________________

Another way to share group marks is to assign a factor to each student, which best represents his or her contribution to the task. The mark factors must total the number of students in the group. The group mark is then multiplied by this factor to arrive at each student's individual mark.

Activity:____________________

Group Mark: 8/10

Number in Group: 4

Mark Factors and Individual Marks

Student's Name	Mark Factor	Individual Mark	Signature
Ahmed	1.0	8/10	____________________
Jasmin	1.0	8/10	____________________
Mike	0.9	7.2/10	____________________
Tabitha	1.1	8.8/10	____________________
Total Mark Factor	4		

In either case, students must sign to show that they are in agreement with the way the individual marks were assigned.

Assessment Rubric

You may also wish to provide students with an **Assessment Rubric** similar to the one shown. Students can use this rubric to assess the manner in which the group worked together.

Assessment Rubric for Group Work: Individual Assessment of the Group

Individual's name:____________________

Names of group members: ____________________

Name of activity: ____________________

Circle the appropriate number: #1 is excellent, #2 is good, #3 is average, and #4 is poor.

1. The group worked cooperatively. Everyone assumed a role and carried it out.	1	2	3	4
2. Everyone contributed to the discussion. Everyone's opinion was valued.	1	2	3	4
3. Everyone assumed the roles assigned to them.	1	2	3	4
4. The group was organized. Materials were gathered, distributed, and collected.	1	2	3	4
5. Problems were addressed as a group.	1	2	3	4
6. All parts of the task were completed within the time assigned.	1	2	3	4

Comments:

If you were to repeat the activity, what things would you change?

Reading Strategies and Science

Coordinated Science for the ***21st Century*** is an activity-driven program that gives reading a high level of relevance. Reading passages are frequently presented within the context of the real-world, motivational activities. Even reluctant readers find a new-found reason to read.

The Reading Process in the Science Classroom

The *Active Learning Instructional Model* is an activity-driven project based program that naturally lends itself to the established strategies for reading in the science content area. The curriculum was developed with the mandate and mission of "science for all students" and the text therefore has many of these literacy strategies embedded in the instructional model. Educators now understand that linking science and language education strengthens students' skills in both areas; the two disciplines are inherently interdependent. The natural synergy between language and science makes the purposeful linking of the two disciplines extremely valuable, especially when using a contextual guided inquiry approach. Science educators want students to be able to think scientifically, understand the reading and vocabulary of the science content presented, as well to as express themselves effectively. The structure of the ***Active Learning*** chapter format is based on the most current knowledge and research on cognitive learning and reading in the science content area.

The following is a summary of many of the specific strategies and techniques embedded in the *Active Learning Instructional Model* that will help teachers fuse science and language experiences in the classroom, and thus help increase student learning and performance in both content areas.

The reading process includes three essential phases: before reading, during reading, and after reading. Each of these phases is implicitly embedded into the *Active Learning Instructional Model.*

Before Reading

- Purpose Setting
- Eliciting Prior Knowledge
- Previewing
- Predicting

During Reading

- Monitoring Comprehension
- Using Context Clues
- Using Text Structure Clues

After Reading

- Reflecting
- Summarizing
- Seeking Additional Information

Before Reading

Purpose Setting – Teachers are asked to introduce each new chapter by whole class readings and discussions on the ***Chapter Scenario***, ***Chapter Challenge*** and ***Criteria*** to let students in on the big picture of their learning. These three beginning components of each chapter were developed to provide relevant background context, set the stage and motivation for the learning and then encourage students to become active participants in the assessing of that knowledge. The ***Chapter Scenario*** and ***Challenge*** addresses real issues relevant to students' lives and provide the purpose for the learning and reading that will follow. The ***Criteria*** for assessing the success of the ***Chapter Challenge*** must also be determined during the class discussion in which the students collectively develop a rubric. This component of the curriculum is geared to let students in on – indeed help create – the criteria by which their challenge will be judged. When students agree to the rubric by which they will be measured, research has shown that the students will perform better and achieve more.

Seven to ten ***Activities*** (inquiry-based hands-on labs) then follow. Each ***Activity*** takes from one to two days to complete and will help the students develop the science content needed to complete their ***Chapter Challenge.***

Eliciting Prior Knowledge – A ***What Do You Think?*** question begins each ***Activity,*** as a means to uncover preconceived ideas and prior knowledge to help students in their learning about the specific content they will be exposed to in the *Activity.* These tasks ask students to think about and then write in their journals their thoughts about the question for a few minutes prior to beginning an *Activity.* The ***What Do You Think?*** question gives students the chance to verbalize what they think about the topic at hand. The questions are designed to elicit both common conceptions and misconceptions. After completing the *Activity* they can then review their prior ideas to see what new understandings on the subject they may have gained.

During Reading

For You To Do or ***Investigate*** is the guided inquiry-based lab in each *Activity*. Students are asked to work in groups as they perform these labs. Within their groups students are asked to read the instructions, perform the tasks requested, and then to record and write up their results and data in logs or lab books. Reading in a group serves to provide support that is absent in independent reading. The collaborative group process helps to ensure that all types of learners participate in the activity and learning process. At this stage of the learning cycle, students do not read about what others have done in science; they read together in a group in order to **do** the science.

The ***For You To Do*** component of the curriculum requires students to use key parts of the During Reading process.

> *Focus attention* – Students must follow the step-by-step instructions, thus they are asked to carefully focus their attention on each step of the process.
>
> *Monitor comprehension* – As they perform the guided steps of the inquiry lab, students must attend to their own comprehension. In reading what to do, you must learn what to do.
>
> *Anticipate and predict* – Following the step-by-step sequence, students naturally begin to question: "What's next?"
>
> *Use text-structure* – The design, with lots of supportive graphics and text-structure clues, gives practice in using all the information at hand.

Short condensed readings (***For You To Read***, or ***Science Talk***, or **Digging Deeper**) follows the lab activities. *Activity Before Content (ABC)* is a central philosophical tenet of the *Active Learning Instructional Model.* These content readings

summarize the science principles learned in the inquiry. These readings extend the learning process by allowing students to integrate new information into what they already have learned.
In these sections science vocabulary may be defined and explained and mathematical equations are presented where appropriate. Insights increase as students use reading to extend their first-hand knowledge of science.

After Reading

Each *Activity* concludes with ***Reflecting on the Activity and the Challenge***, which relates the activity the students have just completed to the bigger picture, their final chapter project. If students have lost sight of the bigger picture while completing the activities, the larger context of the ***Chapter Challenge*** is brought back into sight. In the section ***Preparing for the Chapter Challenge***, students see how each activity forms a piece of the completed puzzle that will become their ***Chapter Challenge.*** In completing the ***Chapter Challenge***, each group will put together a public presentation of the content learned in the chapter in a new real–world context. The students must transfer the content learned from the activities into a new domain. The ***Chapter Challenge*** thus reinforces the readings and science learned from their activities. Each group's presentation also serves as an excellent vehicle for multiple exposures to content, as it repeatedly summarizes the content learned in the chapter.

Assessment Strategies and Opportunities

The word that most aptly describes the assessment opportunities in *Coordinated Science for the **21st Century*** is "authentic." At the conclusion of each chapter, students demonstrate the usefulness of their newly-acquired knowledge by creating a culminating project, the **Chapter Challenge**.

Classroom Assessment and the NSES Assessment Characteristics

In keeping with the discussion on assessment as outlined in the *National Science Education Standards* (NSES), four issues, which may present somewhat new considerations for teachers and students, are of particular importance.

1. Formative and Integrated

The *National Science Education Standards* (NSES) indicates that assessments should be seen as the ongoing process of gathering and analyzing evidence to determine appropriate instruction. That instruction and any assessment must be consistent in format and intent. In addition, students should also be involved in the assessment process.

2. Promoting Deep Learning

Another explicit focus of NSES is to promote a shift to deeper instruction on a smaller set of core science concepts and principles. Assessment can support or undermine that intent. It can support it by raising the priority of in-depth treatment of concepts, so that students recognize the relevance of core concepts. Assessment can undermine a deep treatment of concepts by encouraging students to parrot back large bodies of knowledge-level facts that are not related to any specific context in particular. In short, by focusing on a few concepts and principles, deemed to be of particularly fundamental importance, assessment can help to overcome a bias toward superficial learning. *This is an area that some students will find unusual, if their prior science instruction has led them to rely largely on memorization skills for success.*

3. Flexible and Celebrating Diversity

Students differ in many ways. Assessment that calls on students to give thoughtful responses must allow for those differences. Some students may initially find the open-ended character of *Chapter Challenge* disquieting. They may ask many questions to try to find out exactly what the finished product should look like. Teachers will have to give a consistent and repeated message to those students, expressed in many different ways, that this is an opportunity for students to be creative and to show what they know in a way that makes sense to them. This allows for the assessments to be adapted to students with differing abilities and diverse backgrounds.

4. Contextual

While the *Chapter Challenges* are intended to be flexible, they are also intended to be consistent with the manner in which instruction takes place. The *Active-Learning Instructional Model* is such that students have the opportunity to learn new material in a way that places it in context. Consistent with that, the *Chapter Challenge* calls for the new material to be expressed in context. Traditional tests are less likely to allow this kind of expression, and are more likely to be inconsistent with the manner of teaching that *Coordinated Science for the **21st Century*** is designed to promote.

Authentic Assessment

At the culmination of each chapter, students demonstrate the usefulness of their newly acquired knowledge by completing the *Chapter Challenge*. This authentic and performance-based assessment:

- Focuses on deep understanding of meaningful wholes rather than superficial knowledge of isolated parts
- Encourages the integration of knowledge into the learner's schema rather than the mere reproduction of information discovered by others
- Emphasizes motivational tasks rather than trivial tasks, so is seen to be meaningful in nonschool environments
- Promotes transfer of knowledge to real-life contexts.

Chapter Challenge

The ***Chapter Challenge*** is the cornerstone of *Coordinated Science for the **21st Century***. The challenge provides the purpose for all the learning that takes place. It also provides the central assessment tool for the learning. Students demonstrate their understandings of the content learned in the chapter by creating a group project that they publicly present that ties together the science knowledge gained in the activities. This performance-based assessment promotes: motivation, transfer of knowledge, and the holistic integration of the content learned.

Opportunities to Meet Different Needs – Another great aspect of the ***Chapter Challenge*** lies in the wide variety of tasks that are needed to complete the project. These tasks give students with different talents opportunity to excel. Students who express themselves artistically will have an opportunity to shine in some parts of the challenge. Students who can design and build may take the lead in another part. Some challenges have a major component devoted to writing, while others require oral or visual presentations. All of the challenges require what's important: the demonstration of solid science understanding.

Multiple Exposures to Content – One of the main strengths of this project-based assessment is that students don't just get to create their own project, they get to view every other groups' project. Through this repeated public display, students get what many need – multiple exposures to content. This content has been created by their peers and is thus not seen as "boring" review, but as a chance to view other's creations.

Opportunities for Creativity – In many science courses, all students are expected to converge on the identical solution. In *Coordinated Science for the **21st Century***, each group is expected to create a unique ***Chapter Challenge***. Each group project must demonstrate correct science concepts, but there is ample room for creativity on the students' part.

Chapter Challenge Assessment Rubrics

Student Participation – The discussion of the criteria for grading the project and the creation of a grading rubric is an essential ingredient for student success. After the introduction of the challenge, students get to create their own class grading criteria. "What does an "A" look like?" "Should creativity be weighted more than delivery?" "How many scientific principles need to be included?" The criteria will be visited again at the end of the chapter, but at this point it provides the expectation level that students set for themselves.

Sample Assessment Rubrics – To give all involved ideas for what might be included in the class rubric, at least two sample rubrics are included at the beginning of each chapter in the Teacher Edition. Just as every ***Chapter Challenge*** is different, every suggested rubric emphasizes specific criteria appropriate for that challenge. There are even samples of three different types of rubrics. There are rubrics to evaluate: 1) the Group-Process Work, 2) the Scientific and Technological Thinking, and 3) Performance-Assessment Rubrics to evaluate the actual data-collection methods used within the investigations.

Multiple Assessment Tools and Opportunities

Although the cornerstone of the program is the ***Chapter Challenge***, more traditional assessment tools, such as – activity logs, lab reports, and quizzes are also provided for and encouraged. The teacher's lesson plan book for a 21st Century class will look the same as for any other of their science classes, except for the addition of the ***Chapter Challenge*** assessment. Logs, data charts, and journal writing form an integral part of the assessment strategies in this guided-inquiry program. Every activity provides explicit instructions for what students are to record in their logbooks.

Science to Go – This section provides questions to check understanding of key principles in the activity. They are often completed as homework assignments and can serve as a study guide to review what is most important in each activity.

Alternative Chapter Assessment – For traditional written evaluation of key concepts, an alternative assessment is provided for each chapter. This test can serve as a study guide, a homework assignment, or as an additional method of evaluation to supplement the information gained from the ***Chapter Challenge***.

Safety in the Science Classroom

Coordinated Science for the ***21st Century*** is an active, inquiry-based program. It is founded on the belief that science understanding comes best not from hearing or reading but from *doing*. However, with such doing comes the possibility of injury. Safety is a primary concern.

Safety Guidelines

A critical guideline with respect to safety is student ownership. No teacher can take sole responsibility for making each student safe at all times. The students themselves must also own the responsibility for safety at each and every moment.

Therefore, one primary goal is to ellicit student involvement in safety procedures. Begin a discussion of the necessary rules within your classroom. Solicit student feedback. You might even post these agreed-upon rules for all to see and revisit them periodically.

In addition, have students read the Safety Guidelines provided at the front of their student editions. After reading the rules, students and parents should sign the Safety Contract shown on the next page. It is provided as a blackline master in the Teacher Resources book.

Coordinated Science for the 21st Century

Safety in the Science Classroom

Chemistry is a laboratory science. During this course you will be doing many activities in which safety is a factor. To ensure the safety of you and all students, the following safety rules will be followed. You will be responsible for abiding by these rules at all times. After reading the rules, you and a parent or guardian must sign a safety contract acknowledging that you have read and understood the rules and will follow them at all times. (See page xxxix.)

General Rules

1. There will be no running, jumping, pushing, or other behavior considered inappropriate in the science laboratory. You must behave in an orderly and responsible way at all times.
2. Eating, drinking, chewing gum, or applying cosmetics is strictly prohibited.
3. All spills and accidents must be reported to your teacher immediately.
4. You must follow all directions carefully and use only materials and equipment provided by your teacher. Only activities approved by your teacher may be carried out in the chemistry laboratory.
5. No loose, hanging clothing is allowed in the laboratory; long sleeves must be rolled up; bulky jackets, as well as jewelry, must be removed.
6. Never work in the lab unless your teacher or an approved substitute is present.
7. Identify and know the location of a fire extinguisher, fire blanket, emergency shower, eyewash, gas and water shut-offs, and telephone.

Equipment Rules

1. All equipment must be checked out and returned properly.
2. Do not touch any equipment until you are instructed to do so.
3. Do not use glassware that is broken or cracked. Alert your teacher to any glassware that is broken or cracked.

Working with Chemicals

1. Never touch or smell chemical unless specifically instructed to do so by your teacher. Never taste chemicals.
2. Safety goggles must be worn at all times.
3. Carefully read all labels to make sure you are using the correct chemicals and use only the amount of chemicals instructed by your teacher.
4. Keep your hands away from your face and thoroughly wash with soap and water before exiting the classroom.
5. Contact lenses can absorb certain chemicals. Advise your teacher if you wear contact lenses.
6. Never add water to an acid and always add acid slowly to water.
7. Follow your teachers' instructions for the correct disposal of chemicals. Do not dispose of any chemical waste, including paper towels used for chemical spills, in the trash basket or down a sink drain.

Flame Safety

1. Use extreme caution when using any type of flame. Keep your hands, hair, and clothing away from flames.
2. Long hair must be tied back at all times.
3. Keep all flammable materials away from open flames. Some winter jackets are extremely flammable and should be removed before entering the laboratory.
4. Always point the mouth of a test tube away from yourself or any other person when heating a substance.
5. Extinguish the flame as soon as you are finished.
6. Always use heat-resistant gloves when working with an open flame.

Work Area

1. When working in the laboratory all materials should be removed from the workstation except for instructions and data tables. Materials should not be removed from the desktop to the floor as this is a hazard for someone walking with glassware or chemicals.
2. The work area should be kept clean at all times. After completing an activity, wipe down the area.
3. Notify your teacher of any spills immediately so they can be properly taken care of.

Safety Contract

The following contract may be reproduced and must be signed by each student and a parent or guardian before participating in laboratory activities.

I have read **Safety in the Science Classroom** and understand the requirements fully. I recognize that there are risks associated with any science activity and acknowledge my responsibility in minimizing these risks by abiding by the safety rules at all times.

Please list any known medical conditions or allergies:

____

____

____

____

I **do / do not** wear contact lenses. (Circle one)

Emergency phone contact: __________________________

Student signature __________________________________ Date _______

Parent or guardian _________________________________ Date _______

Teacher____ Date _______

Expanding the 5E Model

A proposed 7E model emphasizes "transfer of learning" and the importance of eliciting prior understanding

Arthur Eisenkraft

SOMETIMES A CURRENT MODEL MUST BE AMENDED TO maintain its value after new information, insights, and knowledge have been gathered. Such is now the case with the highly successful 5E learning cycle and instructional model (Bybee 1997). Research on how people learn and the incorporation of that research into lesson plans and curriculum development demands that the 5E model be expanded to a 7E model.

The 5E learning cycle model requires instruction to include the following discrete elements: *engage*, *explore*, *explain*, *elaborate*, and *evaluate*. The proposed 7E model expands the *engage* element into two components—*elicit* and *engage*. Similarly, the 7E model expands the two stages of *elaborate* and *evaluate* into three components— *elaborate*, *evaluate*, and *extend*. The transition from the 5E model to the 7E model is illustrated in Figure 1.

These changes are not suggested to add complexity, but rather to ensure instructors do not omit crucial elements for learning from their lessons while under the incorrect assumption they are meeting the requirements of the learning cycle.

Eliciting prior understandings

Current research in cognitive science has shown that eliciting prior understandings is a necessary component of the learning process. Research also has shown that expert learners are much more adept at the transfer of learning than novices and that practice in the transfer of learning is required in good instruction (Bransford, Brown, and Cocking 2000).

The *engage* component in the 5E model is intended to capture students' attention, get students thinking about the subject matter, raise questions in students' minds, stimulate thinking, and access prior knowledge. For example, teachers may engage students by creating surprise or doubt through a demonstration that shows a piece of steel sinking and a steel toy boat floating. Similarly, a teacher may place an ice cube into a glass of water and have the class observe it float while the same ice cube placed in a second glass of liquid sinks. The corresponding conversation with the students may access their prior learning. The students should have the opportunity to ask and attempt to answer, "Why is it that the toy boat does not sink?"

The *engage* component includes both accessing prior knowledge and generating enthusiasm for the subject matter. Teachers may excite students, get them interested and ready to learn, and believe they are fulfilling the engage phase of the learning cycle, while ignoring the need to find out what prior knowledge students bring to the topic. The importance of *eliciting* prior understandings in ascertaining what students know prior to a lesson is imperative. Recognizing that students construct knowledge from existing knowledge, teachers need to find out what existing knowledge their students possess. Failure to do so may result in students developing concepts very different from the ones the teacher intends (Bransford, Brown, and Cocking 2000).

A straightforward means by which teachers may elicit prior understandings is by framing a "What Do You Think" question at the outset of the lesson as is done consistently in some current curricula. For example, a common physics lesson on seat belts might begin with a question about designing seat belts for a racecar traveling at a high rate of speed (Figure 2, p. xxxiv). "How would they be different from ones available on passenger cars?" Students responding to this question communicate what they know about seat belts and inform themselves, their classmates, and the teacher about their prior conceptions and understandings. There is no need to arrive at consensus or closure at this point. Students do not assume the teacher will tell them the "right" answer. The "What Do You Think" question is intended to begin the conversation.

The proposed expansion of the 5E model does not exchange the *engage* component for the *elicit* component; the *engage* component is still a necessary element in good instruction. The goal is to continue to excite and interest students in whatever ways possible and to identify prior conceptions. Therefore, the *elicit* component should stand alone as a reminder of its importance in learning and constructing meaning.

Explore and explain

The *explore* phase of the learning cycle provides an opportunity for students to observe, record data, isolate variables, design and plan experiments, create graphs, interpret results, develop hypotheses, and organize their findings. Teachers may frame questions, suggest approaches, provide feedback, and assess understandings. An excellent example of teaching a lesson on the metabolic rate of water fleas (Lawson 2001) illustrates the effectiveness of the learning cycle with varying amounts of teacher and learner ownership and control (Gil 2002).

Students are introduced to models, laws, and theories during the *explain* phase of the learning cycle. Students summarize results in terms of these new theories and models. The teacher guides students toward coherent and consistent generalizations, helps students with distinct scientific vocabulary, and provides questions that help students use this vocabulary to explain the results of their explorations. The distinction between the *explore* and *explain* components ensures that concepts precede terminology.

Applying knowledge

The *elaborate* phase of the learning cycle provides an opportunity for students to apply their knowledge to new domains, which may include raising new questions and hypotheses to explore. This phase may also include related numerical problems for students to solve. When students explore the heating curve of water and the related heats of fusion and vaporization, they can then perform a similar experiment with another liquid or, using data from a reference table, compare and contrast materials with respect to freezing and boiling points. A further elaboration may ask students to consider the specific heats of metals in comparison to water and to explain why pizza from the oven remains hot but aluminum foil beneath the pizza cools so rapidly.

The elaboration phase ties directly to the psychological construct called "transfer of learning" (Thorndike 1923). Schools are created and supported with the expectation that more general uses of knowledge will be found outside of school and beyond the school years (Hilgard and Bower 1975). Transfer of learning can range from transfer of one concept to another (e.g., Newton's law of gravitation and Coulomb's law of electrostatics); one school subject to another (e.g., math skills applied in

FIGURE 1

The proposed 7E learning cycle and instructional model.

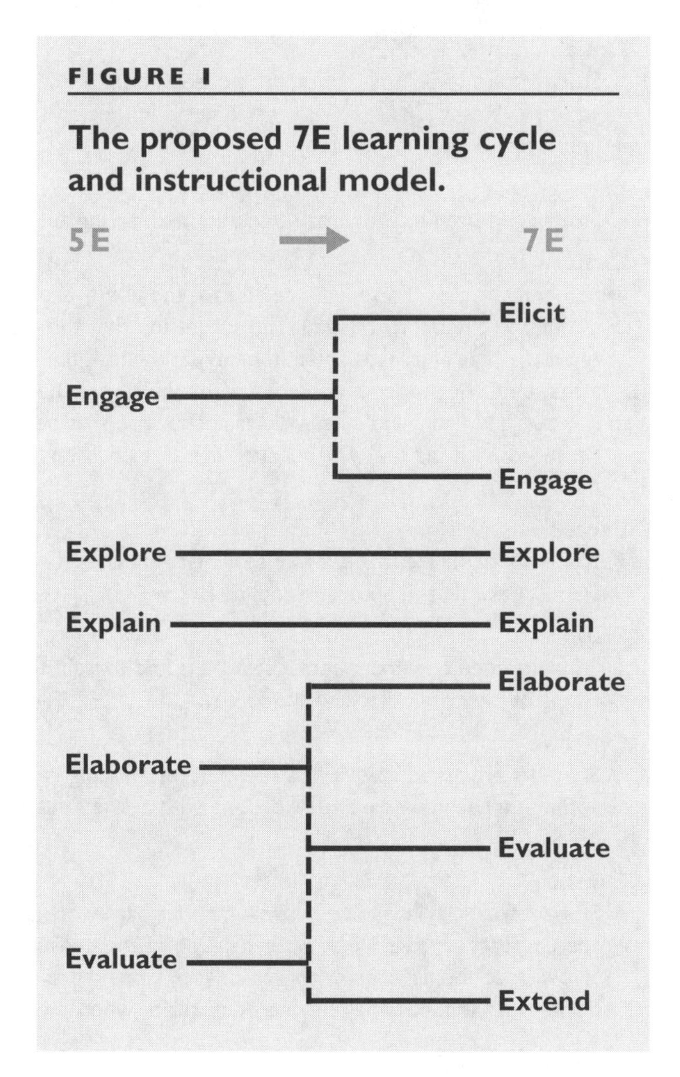

FIGURE 2

Seat belt lesson using the 7E model.

Elicit prior understandings

◆ Students are asked, "Suppose you had to design seat belts for a racecar traveling at high speeds. How would they be different from ones available on passenger cars?" The students are required to write a brief response to this "What Do You Think?" question in their logs and then share with the person sitting next to them. The class then listens to some of the responses. This requires a few minutes of class time.

Engage

◆ Students relate car accidents they have witnessed in movies or in real life.

Explore

◆ The first part of the exploration requires students to construct a clay figure they can sit on a cart. The cart is then crashed into a wall. The clay figure hits the wall.

Explain

◆ Students are given a name for their observations. Newton's first law states, "Objects at rest stay at rest; objects in motion stay in motion unless acted upon by a force."

Engage

◆ Students view videos of crash-test dummies during automobile crashes.

Explore

◆ Students are asked how they could save the clay figure from injury during the crash into the wall. The suggestion that the clay figure will require a seat belt leads to another experiment. A thin wire is used as a seat belt. The students construct a seat belt from the wire and ram the cart and figure into the wall again. The wire seat belt keeps the clay figure from hitting the wall, but the wire slices halfway through the midsection.

Explain

◆ Students recognize that a wider seat belt is needed. The relationship of pressure, force, and area is introduced.

Elaborate

◆ Students then construct better seat belts and explain their value in terms of Newton's first law and forces.

Evaluate

◆ Students are asked to design a seat belt for a racing car that travels at 250 km/h. They compare their designs with actual safety belts used by NASCAR.

Extend

◆ Students are challenged to explore how air bags work and to compare and contrast air bags with seat belts. One of the questions explored is, "How does the air bag get triggered? Why does the air bag not inflate during a small fender-bender but does inflate when the car hits a tree?"

scientific investigations); one year to another (e.g., significant figures, graphing, chemistry concepts in physics); and school to nonschool activities (e.g., using a graph to calculate whether it is cost effective to join a video club or pay a higher rate on rentals) (Bransford, Brown, and Cocking 2000).

Too often, the elaboration phase has come to mean an elaboration of the specific concepts. Teachers may provide the specific heat of a second substance and have students perform identical calculations. This practice in transfer of learning seems limited to near transfer as opposed to far or distant transfer (Mayer 1979). Even though teachers expect wonderful results when they limit themselves to near transfer with large similarities between the original task and the transfer task, they know students often find elaborations difficult. And as difficult as near transfer is for students, the distant transfer is usually a much harder road to traverse. Students who are quite able to discuss phase changes of substances and their related freezing points, melting points, and heats of fusion and vaporization may find it exceedingly difficult to transfer the concept of phase change as a means of explaining traffic congestion.

Practicing the transfer of learning

The addition of the *extend* phase to the *elaborate* phase is intended to explicitly remind teachers of the importance for students to practice the transfer of learning. Teachers need to make sure that knowledge is applied in a new context and is not limited to simple elaboration. For instance, in another common activity students may be required to invent a sport that can be played on the moon. An activity on friction informs students that friction increases with weight.

Because objects weigh less on the moon, frictional forces are expected to be less on the moon. That elaboration is useful. Students must go one step further and extend this friction concept to the unique sports and corresponding play they are developing for the moon environment.

The *evaluate* phase of the learning cycle continues to include both formative and summative evaluations of student learning. If teachers truly value the learning cycle and experiments that students conduct in the classroom, then teachers should be sure to include aspects of these investigations on tests. Tests should include questions from the lab and should ask students questions about the laboratory activities. Students should be asked to interpret data from a lab similar to the one they completed. Students should also be asked to design experiments as part of their assessment (Colburn and Clough 1997).

Formative evaluation should not be limited to a particular phase of the cycle. The cycle should not be linear. Formative evaluation must take place during all interactions with students. The *elicit* phase is a formative evaluation. The *explore* phase and *explain* phase must always be accompanied by techniques whereby the teacher checks for student understanding.

Replacing *elaborate* and *evaluate* with *elaborate*, *extend*, and *evaluate* as shown in Figure 1, p. xxxiii, is a way to emphasize that the transfer of learning, as required in the extend phase, may also be used as part of the evaluation phase in the learning cycle.

Enhancing the instructional model

Adopting a 7E model ensures that eliciting prior understandings and opportunities for transfer of learning are not omitted. With a 7E model, teachers will *engage* and *elicit* and students will *elaborate* and *extend*. This is not the first enhancement of instructional models, nor will it be the last. Readers should not reject the enhancement because they are used to the traditional 5E model, or worse yet, because they hold the 5E model sacred. The 5E model is itself an enhancement of the three-phrase learning cycle that included exploration, invention, and discovery (Karplus and Thier 1967.) In the 5E model, these phases were initially referred to as *explore*, *explain*, and *expand*. In another learning cycle, they are referred to as exploration, term introduction, and concept application (Lawson 1995).

The 5E learning cycle has been shown to be an extremely effective approach to learning (Lawson 1995; Guzzetti et al. 1993). The goal of the 7E learning model is to emphasize the increasing importance of eliciting prior understandings and the extending, or transfer, of concepts. With this new model, teachers should no longer overlook these essential requirements for student learning.

Arthur Eisenkraft is a project director of Active Physics and currently the Distinguished Professor of Science Education and a Senior Research Fellow at the University of Massachusetts, Boston; and a past president of NSTA, e-mail: eisenkraft@worldnet.att.net.

REFERENCES

Bransford, J.D., A.L. Brown, and R.R. Cocking, eds. 2000. *How People Learn*. Washington, D.C.: National Academy Press.

Bybee, R.W. 1997. *Achieving Scientific Literacy*. Portsmouth, N.H.: Heinemann.

Colburn, A., and M.P. Clough. 1997. Implementing the learning cycle. *The Science Teacher* 64(5): 30–33.

Gil, O. 2002. Implications of inquiry curriculum for teaching. Paper presented at National Science Teachers Association Convention, 5–7 December, in Albuquerque, N.M.

Guzzetti B., T.E. Taylor, G.V. Glass, and W.S. Gammas. 1993. Promoting conceptual change in science: A comparative meta-analysis of instructional interventions from reading education and science education. *Reading Research Quarterly* 28: 117–159.

Hilgard, E.R., and G.H. Bower. 1975. *Theories of Learning*. Englewood Cliffs, N.J.: Prentice Hall.

Karplus, R., and H.D. Thier. 1967. *A New Look at Elementary School Science*. Chicago: Rand McNally.

Lawson, A.E. 1995. *Science Teaching and the Development of Thinking*. Belmont, Calif.: Wadsworth.

Lawson, A.E. 2001. Using the learning cycle to teach biology concepts and reasoning patterns. *Journal of Biological Education* 35(4): 165–169.

Mayer, R.E. 1979. Can advance organizers influence meaningful learning? *Review of Educational Research* 49(2): 371–383.

Thorndike, E.L. 1923. *Educational Psychology*, Vol. II: *The Psychology of Learning*. New York: Teachers College, Columbia University.

Reprinted with permission from *The Science Teacher* (70(6): 56–59), a journal for high school science educators published by the National Science Teachers Association (www.nsta.org).

Unit 3

Active Biology™

Chapter 9: A Vote for Ecology
Chapter 10: A Highway Through the Past

A VOTE FOR ECOLOGY
Chapter 9
SAVE THE TREES
SAVE OUR JOBS
We need our jobs!
PROTECT THE ENVIRONMENT
otect our environment

ACTIVE BIOLOGY CHAPTER 9

A VOTE FOR ECOLOGY

Chapter Overview

This chapter provides the teacher with an opportunity to open the eyes of students to the seemingly simple and yet very complex issues in their environment. The chapter starts with a challenge – to create a booklet addressing one current environmental issue, with the primary purpose of educating a targeted audience. The challenge becomes clearer as the students progress through the nine hands-on activities that they do before they complete the **Chapter Challenge.** After learning about diversity in living things in **Activity 1**, the students delve into relationships among organisms in terms of who eats whom found in **Activity 2** and energy flows in **Activity 3**. The students then deal with the factors that affect population size in **Activity 4** followed by competition among organisms in **Activity 5**. In **Activity 6**, the students learn about succession. Succession provides a better understanding of what happened before in the environment, which then leads to the question, "What will happen next?" Since the ecosystem necessarily includes the abiotic component, **Activities 7, 8,** and **9** cover the water cycle, the carbon and oxygen cycle in the processes of photosynthesis and respiration, and the nitrogen and phosphorous cycle, respectively. Each of the nine activities ends with a reflection where students are redirected and refocused to the **Chapter Challenge** found in the very beginning of the chapter.

Chapter Goals For Students

- Identify the roles and importance of consumers, producers, and decomposers in an ecosystem
- Explain how matter is cycled and energy flows through ecosystems
- Provide the meaning and the importance of biodiversity
- Describe how changes in an ecosystem are determined and how they can be analyzed
- Tell how fluctuations in size of a population are determined by birth, death, immigration, and emigration

Chapter Timeline

About one week before beginning the unit: Complete **Activity 5–For You To Do** Part A: Steps 1 – 5.

Day	Activity	Homework
1	Introduce Scenario and Chapter Challenge. Assign groups to respond to the Chapter Challenge. Discuss Criteria and establish a rubric for grading the booklet.	
2	Allow time for observations for Activity 5. Complete rubric for grading the booklet. What Do You Think? questions for Activity 1	Read through steps of For You To Do for Activity 1.
3	Complete all observations for Activity 1, Part A.	Complete Activity 1, Part B.
4	Recapitulate and reflect on Activity 1. Read and discuss BioTalk. Allow time for observations for Activity 5.	Complete Biology to Go questions.
5	Discuss responses to Biology to Go questions. What Do You Think? question for Activity 2. Complete Parts A and B For You to Do.	Read BioTalk and answer Biology to Go questions 1–4.
6	Complete and discuss responses to Biology to Go questions. Recapitulate and reflect on Activity 2. Allow time for observations for Activity 5. What Do You Think? question for Activity 3.	Read through steps of For You To Do for Activity 3.
7	Complete For You to Do.	Read BioTalk.
8	Recapitulate and reflect on Activity 3. Discuss BioTalk. Allow time for observations for Activity 5. Begin answering Biology to Go questions.	Complete Biology to Go questions.
9	Discuss responses to Biology to Go questions. What Do You Think? question for Activity 4. Begin For You To Do.	Read BioTalk.
10	Complete For You to Do. Discuss BioTalk. Recapitulate and reflect on Activity 4. Allow time for observations for Activity 5.	Complete Biology to Go questions. (You may wish to leave the answer for question 5 for the following class.)
11	Discuss responses to Biology to Go questions. What Do You Think? questions for Activity 5. Recapitulate and reflect on Activity 5.	Read BioTalk and complete Biology to Go questions 1–4.
12	Discuss BioTalk. Discuss responses to Biology to Go questions 1–4. Allow class time to discuss question 5. Complete steps 1–6 of Activity 9.	Answer Biology to Go question 5.

Day	Activity	Homework
13	What Do You Think? questions for Activity 6. Complete Activity 6, For You To Do. Allow time for observations for Activity 9.	Read BioTalk and complete Biology to Go.
14	Discuss BioTalk. Discuss responses to Biology to Go. Recapitulate and reflect on Activity 6. Allow time for observations for Activity 9.(Place plants for Activity 8 in the dark.)	Read steps of For You To Do.
15	What Do You Think? questions for Activity 7. Complete Activity 7, For You To Do. Allow time for observations for Activity 9.	Read BioTalk.
16	Discuss BioTalk. Answer and discuss Biology to Go questions. Recapitulate and reflect on Activity 7. Allow time for observations for Activity 9.	Read steps of For You To Do.
17	What Do You Think? question for Activity 8. Complete Day 1 of For You To Do. Allow time for observations for Activity 9.	
18	Complete Day 2 of For You To Do. Allow time for observations for Activity 9.	
19	Allow class time for students to begin preparing for the Chapter Challenge. Allow time for observations for Activity 9.	
20	Complete Day 4 of For You To Do. Allow time for observations for Activity 9.	Read BioTalk.
21	Discuss Bio Talk. Begin answering Biology to Go questions. Allow time for observations for Activity 9.	Complete Biology to Go questions.
22	Discuss responses to Biology to Go. Recapitulate and reflect on Activity 8. Allow time for observations for Activity 9.	
23	What Do You Think? question for Activity 9. Complete remaining steps of For You To Do, Activity 9.	Read BioTalk.
24	Discuss BioTalk. Begin answering Biology to Go questions.	Complete Biology to Go questions.
25	Discuss responses to Biology to Go. Recapitulate and reflect on Activity 9. Review and revise the rubric for the Chapter Challenge, if necessary.	
26	Prepare to respond to the Chapter Challenge.	Prepare to respond to the Chapter Challenge.
27	Prepare to respond to the Chapter Challenge.	Prepare to respond to the Chapter Challenge.
28	Presentation and assessment	Prepare to respond to the Chapter Challenge.

NOTES

National Science Education Standards

Creating a booklet addressing one current environmental issue for the public to be better able to decide how to vote on any given issue is the **Scenario** in this chapter. As students explore diversity in living things, dependence of the organisms on one another, the factors that affect population size, competition among organisms, succession, the water cycle, the carbon and oxygen cycle and their role in the processes of photosynthesis and respiration, and nitrogen and phosphorous cycle, they begin to develop the content understandings outlined below.

CONTENT STANDARDS

Unifying Concepts and Processes

- Systems, order, and organization
- Evidence, models, and explanation
- Change, constancy, and measurement
- Form and function

Science as Inquiry

- Develop abilities necessary to do scientific inquiry
- Identify questions and concepts that guide scientific investigations
- Design and conduct scientific investigations
- Use technology and mathematics to improve investigations and communications
- Formulate and revise scientific explanations and models using logic and evidence
- Recognize and analyze alternative explanations and models
- Communicate and defend a scientific argument
- Develop understandings about scientific inquiry

Life Science

- Interdependence of organisms
- Matter, energy, and organization in living systems
- Behavior of organisms

Science and Technology

- Develop abilities of technological design
- Propose designs and choose between alternative solutions
- Communicate the problem, process, and solution
- Develop understandings about science and technology

Science in Personal and Social Perspectives

- Personal and community health
- Population growth
- Natural resources
- Environmental quality
- Natural and human-induced hazards
- Science and technology in local, national, and global challenges

History and Nature of Science

- Science as a human endeavor
- Nature of scientific knowledge
- Historical perspectives

Key Science Concepts and Skills

Activities Summaries	Biology Principles
Activity 1: Diversity in Living Things Students observe a variety of living organisms, as well as preserved slides to introduce them to the incredible diversity in the animal kingdom and to examine the relationship between structure and function.	• Structure determines the function • Biodiversity is important to the stability of an ecosystem
Activity 2: Who Eats Whom? This activity introduces students to food chains and food webs to gain a sense of how dependent organisms are to each other.	• Food chains and food webs • Producers, consumers, and decomposers • Autotrophs and heterotrophs • Herbivores, carnivores, and omnivores
Activity 3: Energy Flow in Ecosystems Students record and graph the temperatures in three containers of water. One container is left alone while a student puts a moving hand in the second container and a still hand in the third. Students infer from the data that heat is dissipated from the body.	• Laws of thermodynamics • Transfer of energy in a food chain • Energy is lost in the form of heat • Pyramids of biomass, of numbers, and of energy • Trophic levels
Activity 4: Factors Affecting Population Size Students compare the growth of a reindeer population on an island near Alaska and the human population. They examine the roles played by carrying capacity, doubling time, and the four rates that determine a population's size – birthrate, death rate, immigration, and emigration.	• Carrying capacity • Doubling time • Birthrate, death rate, immigration, and emigration • Open and closed populations
Activity 5: Competition Among Organisms Students observe the difference in the growth of plants when they must compete for space and nutrients. They also observe what happens when different species compete. They then design and carry out an experiment to study competition under different conditions.	• Competition • Nonnative species • Invasive species
Activity 6: Succession in Communities Students do a visual scrutiny of a diagram on the changes that occurred along the coastal areas and in the inland areas after two volcanoes erupted on the island of Krakatoa. This activity hones the students' observational skills as they investigate the sequence of changes that take place in succession.	• Primary and secondary succession • Pioneer community • Seral communities • Climax community

Key Science Concepts and Skills	
Activities Summaries	**Biology Principles**
Activity 7: The Water Cycle Students conduct an experiment to measure the rate of transpiration as they learn about the water cycle. They also investigate factors that could affect experimental results as well as possible experimental errors.	• Water cycle • Evaporation, condensation, precipitation, transpiration
Activity 8: Photosynthesis, Respiration, and the Carbon Cycle This activity shows the role of sunlight in photosynthesis. With the use of the chemical iodine, the students are allowed to figure out whether or not starch is produced. This would lead them to answer whether or not photosynthesis occurred. Using petroleum jelly, they are encouraged to examine what part of the leaf allows for gas exchange – (carbon dioxide is taken in and oxygen is released) which is crucial for photosynthesis and respiration to happen.	• Photosynthesis • Respiration • Carbon cycle • Oxygen cycle
Activity 9: The Nitrogen and Phosphorous Cycles Students conduct an experiment to measure the effect of lawn fertilizer and detergent on the growth of algae.	• Nitrogen cycle • Phosphorous cycle • Nitrogen fixation • Denitrification

Equipment List for Chapter 9

Materials needed for each group per activity.

Activity 1 (per group)

Station 1

- stereomicroscope or 10x hand lens
- watch glass
- dropping pipette
- compound microscope
- no. 2 paintbrush
- prepared slide of longitudinal section of hydra
- 1 or 2 live hydra
- daphnia culture

Station 2

- stereomicroscope or 10x hand lens
- watch glass
- 1 or 2 live planaria
- small piece of raw liver
- no. 2 paintbrush
- compound microscope
- prepared slide of cross section of planaria

Station 3

- hand lens
- paper towels (moistened)
- compound microscope
- 1 or 2 live earthworms
- box containing damp soil
- prepared slide of cross section of earthworm

Station 4

- no. 2 paintbrush
- shallow box or pan
- small pieces of fruit or lettuce
- 1 live land hermit crab

Station 5

- aquarium
- live frog

Activity 2

- ball of string about 35 m long (one for class)
- plastic name tag holders (one per student)

Activity 3 (per group)

- 3 large containers (1000 L)
- 3 thermometers
- ice
- stirring rod

(per student)

- graph paper

Activity 4 (per group)

- calculator

(per student)

- graph paper

Activity 5 (per group)

- 6 milk container bases (or similar containers)
- potting soil
- cress seeds (about 80)
- lettuce seeds (about 15)
- variety of other seeds (students' choice)
- small watering can (or suitable substitute)
- ruler

Activity 6

- no materials are needed

Activity 7 (per group)

- 40 cm piece of plastic tubing
- 0.1 mL pipette
- water tank or wide-mouthed jar made of plastic or glass
- branch of Coleus or Zebrina
- sharp knife or garden scissors
- paper towels
- iron clamp
- timer
- graphing papers
- ruler
- calculator

Activity 8 (per class)

- microwave
- Erlenmeyer flasks

Day 1 (per group)

- 2 4-inch potted plant
- 1 pair of scissors
- 1 beaker
- 1 baby food jar with lid
- 2 forceps
- isopropyl alcohol
- hot (60° C) tap water
- petroleum jelly
- thermometer
- 4 sets of gloves
- 4 goggles
- 4 aprons
- soap
- paper towels

Day 2

- the baby food jar with the isopropyl alcohol from **Day 1**
- 1 baby food jar
- the same forceps used for the isopropyl alcohol from **Day 1**
- 1 forceps
- 2 Petri dishes
- Lugol's iodine solution
- white paper
- 4 sets of gloves
- 4 goggles
- 4 aprons
- soap
- paper towels

Day 4

- the same 4-inch potted plant originally kept in the dark and then was exposed to sunlight from **Day 1**
- Histoclear
- 8 swab applicators
- paper towels
- 1 baby food jar
- Lugol's iodine solution
- 4 Petri dishes
- 4 white papers
- 4 sets of gloves
- 4 goggles
- 4 aprons
- soap

Activity 9 (per group)

- 3 1-L jars
- 2.25 L distilled water (750 mL per jar
- 30 mL pond water with algae (10 mL per jar)
- 45 gm lawn fertilizer (15 gm in 1 jar)
- 45 gm detergent (15 gm in 1 jar)
- plastic wrap
- glass marker
- 3 filter papers
- 3 funnels
- scale
- gloves
- goggles
- apron
- soap
- paper towels

NOTES

A Vote for Ecology

Scenario

Many Americans have begun to realize the importance of ecological issues. Americans now care about problems that were almost unknown a few years ago. Land and water management, pollution, biodiversity, invasive species, and many more concerns are on people's minds.

However, ecological issues cannot be considered on their own. They must be included in the economic and social spheres. The hidden costs of environmental programs are sometimes forgotten. It is necessary to develop a balanced solution to problems. This is the only reasonable way to sustain the environment.

For example, fishing provides food, income, and employment for millions of people. However, fishing has environmental costs. Rare species may be threatened. Marine ecosystems can be disturbed. Also, it is questionable how long the resource will last. Aquaculture presently offers an alternative. It provides a chance to expand the food supply from freshwater or the sea. However, aquaculture can also be ecologically unsound. Natural habitats are lost. The introduction of alien species in an area can pose a threat to the existing ecosystems. The spread of disease from farmed to wild populations is also a concern.

The League of Concerned Voters in your area recognizes the importance of preserving the environment. However, they are also aware that a lack of information about ecological issues could lead to conflict rather than constructive action. That is why they have decided to commission the development of a series of booklets. These booklets are intended to introduce the scientific facts behind current issues.

Scenario and Chapter Challenge

The scenario described leads to the focal point of this chapter, which is for the students to make a booklet on one current environmental issue with the purpose of educating a targeted audience. This topic will not be new to the students and they will be very excited. To get them focused, emphasize that there will be nine activities to be done and that they all contribute to a more thorough understanding of the task. Explain that after each activity, they will be made to reflect on what they have done with one goal in mind – how does the activity help in the resolution of the **Chapter Challenge**?

You may wish to show students booklets about environmental issues, if they are available. Also, tell your students that they will be given enough time for the **Chapter Challenge**. Remind them, however, that procrastination will not work. Entertain as many questions that the students will ask. It is only through their questions that you will have a better perception of their grasp of the task ahead of them.

Assign the steps required in the book to teams. Give them time to discuss it among themselves and then ask them to explain it to the class. Ask the class if they can add to the explanations already given. Allow them to state their disagreement for as long as they can explain why they do so.

Chapter Challenge

Your challenge is to create a booklet addressing one current issue. These booklets will be provided to the public. The League hopes that this will produce an informed public, who is better able to decide how to vote on any given issue. Before you begin, you will need to decide which audience you are targeting. You may choose to write your booklet geared to adult voters, teenagers, or a child. Regardless of which you choose, it is assumed that the readers will be non-specialists and the text should be written with this in mind. The booklets should be easily understood by your target audience.

In producing your booklet, you should:

- identify and research one current issue that threats the environment
- provide the relevant data on the issue
- draw attention to areas where data may be weak or lacking
- interpret the data and indicate the limits of the interpretation.

In providing the science behind the issue, you should:

- identify the roles and importance of consumers, producers, and decomposers in an ecosystem
- explain how matter is cycled and energy flows through ecosystems
- provide the meaning and the importance of biodiversity
- describe how changes in an ecosystem are determined and how they can be analyzed
- tell how fluctuations in size of a population are determined by birth, death, immigration, and emigration.

Criteria

How will your booklet be graded? What qualities should a good booklet have? Discuss these matters with your small group and with your class. You may decide some or all of the following qualities are important:

- significance of the issue identified
- completeness and accuracy of the ecology principles presented
- merit of the interpretations suggested
- readability of the booklet
- design and layout of the booklet.

You may have additional qualities that you would like to include. Once you have determined the criteria that you wish to use, you will need to decide on how many points should be given to each criterion. Your teacher may wish to provide you with a sample rubric to help you get started.

531

Criteria

One way for the students to have ownership of their learning process is for them to be involved in the assessment as well. The best way to do this is to ask for their input in defining the criteria to be used for their rubric. Also, students should provide input into how many points or how much weight should be given for each element of the **Chapter Challenge**. Tell them that with their rubric, they can target the grade that they want to get. Explain to them with the help of their rubric, they will actually be grading the booklets of the other groups.

An example of a rubric is shown on the following page. You can use it as is or modify it, taking into consideration your state science content Standards, your particular assessment style, and your students' ideas.

Assessment Rubric for Chapter Challenge

	4	3	2	1
Informative Identifies the roles and importance of consumers, producers, and decomposers in an ecosystem. Explains how matter is cycled and energy flows through ecosystems. Provides the meaning and importance of biodiversity. Describes how changes in an ecosystem are determined and how they can be analyzed. Tells how fluctuations in size of a population are determined by birth, death, immigration, and emigration.	All five are correct	Only four are correct	Only three are correct	Only two are correct
Creative Color Original Catches attention	All three are present	Only two are present	Only one is present	None is present
Grammar Grammatically correct Complete sentences used Punctuations are correct	All three are present	Only two are present	Only one is present	None is present
Resources Internet Periodicals Books	All three are present	Only two are present	Only one is present	None is present

NOTES

ACTIVITY 1– DIVERSITY IN LIVING THINGS

Background Information

The Biosphere

Spread out over the Earth's surface, between the solid rocky crust of the inner Earth, and the upper reaches of the atmosphere, extends the world of life–a film of living matter. It is made up of grass, shrubs, and trees, of micro-organisms, of worms, fish, rabbits, and wolves. This thin layer of life, along with the air, soil, water, and other nonliving matter that surrounds it, is called the biosphere.

Organisms exist that live in almost every conceivable environment. There are algae that grow only on melting snow. Many plants and animals spend their entire lives in lightless caves. Some algae and bacteria live in hot springs where temperatures are near the boiling point. That all these habitats are populated is evidence of genetic variability and natural selection.

That is the focus of this chapter of *Coordinated Science for the 21st Century*, and the first activity introduces students to the incredible biodiversity of the animal kingdom. This activity is not an attempt to walk the students through the animal phyla. Its purpose is to arouse the students' interest in the diversity of life, develop their observational skills, and perhaps to encourage them to generate questions that they may wish to pursue in other biology courses.

Biodiversity in the Animal Kingdom

The animal kingdom presents an enormous diversity of organisms. Although they display many structural and functional differences, all animals share certain characteristics. All are essentially multicellular in organization, and heterotrophic. Most are capable of locomotion for some part of their lives, and most are able to respond rather quickly to changes in their environment.

Scientists divide this enormously complex and large kingdom into two large groups: the invertebrates and the vertebrates. The invertebrates do not have a backbone, and the vertebrates have a "backbone" (notochord) for at least some time in their lives. Students will be asked to consider if the organisms they are observing are invertebrates or vertebrates. Students may also find it interesting to realize that over 95% of the described animal species are invertebrates!

Complexity and evolutionary development can also be looked at from the point of view of the body cavity. Although the body cavity, called the coelom, is important in classifying animals, it is not within the scope of this biology course.

Zygote (the cell resulting from the union of male and female sex cells) development provides another way of dividing animals into phyla, but is also not examined in this activity.

Symmetry also provides clues to the complexity and evolutionary development of an organism. More complex animals display bilateral symmetry. Students will have an opportunity to view organisms that display both radial and bilateral symmetry.

The simplest organism that the students will encounter in this activity is the hydra, from the phylum Cnidaria. (Polifera, or the sponges, are even less complex animals, and at one time were considered to be plants.) The Cnidaria include hydra, jellyfish, and coral. Members of this phylum, which consists of about 10,000 species, are found only in aquatic ecosystems. Cnidarians have only two germ layers, the ectoderm and the endoderm. However, unlike sponges, they have true tissues: nerve, muscle, and digestive.

The planarian, a member of the phylum Platyhelminthes, or the flatworms, represent a further increase in complexity. Planaria show bilateral symmetry and have a primitive brain. They also have true organs and rudimentary organ systems for digestion and excretion.

The phylum Annelida, or the segmented worms, possess a true coelom. Segmentation is also an important evolutionary advantage that annelids share with humans. Segmentation permits greater specialization. The earthworm is an annelid from the class Oligochaeta. Other annelids include leeches and bristled marine worms.

Arthropods are the pinnacle of adaptation and diversity. They are a large group of animals that can be found in all habitats. They include: spiders, scorpions, ticks, mites, lobsters, crabs, crayfish, centipedes, millipedes, and 750,000 species of insects like beetles, flies, ants, bees, wasps, moths, and butterflies. The key feature separating arthropods from other species is the exoskeleton. More complex arthropods show an increase in brain size and complexity, and thus an increase in sensory abilities. Students will note that hermit crabs have eyes and antennae with which they sense their environment.

Unlike invertebrates, the chordates are organisms that at some point in their development have a stiff rod, the notochord, running down their back. About 95% of chordates belong to the subphylum Vertebrata. Vertebrates have a hollow, bony structure that surrounds the dorsal nerve cord. They also have an endoskeleton, a large brain protected by a skull, a complex heart and circulatory system, an advanced nervous system, and a large coelom that contains the vital organs. Frogs are amphibians. They are born in fresh water and live initially as gilled tadpoles. They then change into the adult form that the students will be observing. The adult is air-breathing and lives on land.

Goals and Assessments

Goal	Location in Activity	Assessment Opportunity
Observe a group of diverse organisms.	**For You To Do** Part A	Student observations are similar to those listed in the Teacher's Edition.
Relate the structure of an organism to its adaptation to the environment.	**For You To Do** Part B	**For You To Do Part B** (a) Students are able to relate structure of the organism to two different functions. **Biology To Go** Question 6.
Describe the organization of the biosphere.	**BioTalk**	**Biology to Go** Question 1.
Define biodiversity and explain its importance.	**BioTalk**	**Biology to Go** Questions 2 and 4.
Explain the effect of human activity on biodiversity.	**BioTalk**	**Biology to Go** Question 3.
Read about the effects of extinction.	**BioTalk**	**Inquiring Further** Questions 1 and 2.
Practice safe laboratory techniques for handling living organisms.	**For You To Do** Part A	Students follow the safety directions provided. Table, equipment, and hands are carefully washed.

Activity Overview

This activity introduces students to the concept of biodiversity. They observe a variety of living organisms, as well as preserved slides. They are given the opportunity to look at hydra, planarians, earthworms, hermit crabs, and frogs. This investigation is not intended to be an abbreviated study of animal phyla. Its primary purpose is to sharpen students' observations of living animals and introduce them to the incredible diversity in the animal kingdom. Students will also examine the relationship between structure and function through observing the adaptations of different organisms to their environment.

Preparation and Materials Needed

Preparation

You will need to ensure before beginning this unit that the living organisms required for this activity are available for your class. This is an exciting, inviting, hands-on approach for students to begin their study of biology. It is definitely worth the preparation time to engage the students in this way. Preserved as well as live specimens can be purchased from biological supply houses.

The stations will need to be set up before the students arrive for class.

You may wish to duplicate **Blackline Master Ecology 1.1: Comparing Animals** data sheet for the students to use to record their observations.

Materials/Equipment Needed (per group)

Station 1

- stereomicroscope or 10x hand lens
- watch glass
- dropping pipette
- compound microscope
- no. 2 paintbrush
- prepared slide of longitudinal section of hydra
- 1 or 2 live hydra
- daphnia culture

Station 2

- stereomicroscope or 10x hand lens
- watch glass
- 1 or 2 live planaria
- small piece of raw liver
- no. 2 paint brush
- compound microscope
- prepared slide of cross section of planaria

Station 3

- hand lens
- paper towels (moistened)
- compound microscope
- 1 or 2 live earthworms
- box containing damp soil
- prepared slide of cross section of earthworm

Station 4

- no. 2 paintbrush
- shallow box or pan
- small pieces of fruit or lettuce
- 1 live land hermit crab

Station 5

- aquarium
- live frog

Learning Strategies for Students with Limited English Proficiency

1. In the **Investigate** section students will encounter vocabulary they must understand to complete their observations. New vocabulary words may include:

characteristics	habitat	symmetry	skeleton
cross section	appendages	oxygen	endoskeleton
exoskeleton	(un)segmented	digestive cavity	longitudinal section

   Provide all students with an opportunity to observe the living organisms using primarily their natural curiosity and observational skills. As the students make their observations, encourage them to connect what they see to the new vocabulary words. Keep in mind that the objective of the activity is to sharpen students' observational skills and to gain an appreciation for the diversity of life. The vocabulary is there to guide students in their observations. Substitute vocabulary words or use diagrams, as appropriate for your class.

2. Point out new vocabulary words in context of the **BioTalk**. Practice using the words as much as possible.

biosphere	biotic	organism	population
community	abiotic	ecosystem	extinction
species	biodiversity		

   You may wish to relate the ideas of an organism, population, and community to the students' immediate experiences.

3. Consider using a CLOZE activity at the end of this activity. CLOZE activities are useful tools for summarizing material and giving English-language learners opportunities to practice writing complete sentences using science vocabulary. CLOZE activities are most effective when used frequently to build students' ability with more complex sentences. Ask students to describe the organization in the biosphere. You may wish to offer a first sentence as an example. For instance, "The biosphere is where living things are found." Model the writing process of editing, where students make corrections that improve the sentences.

Once the paragraph is complete and students agree that it accurately summarizes the organization of the biosphere, have them copy down the paragraph. Explain that tomorrow there will be a brief quiz on the paragraph. The quiz will include the same paragraph they wrote down, but with several blanks with words that are missing. Prepare the quiz by keying in the paragraph and then going back and removing every fifth word and replacing it with a blank. Score the quiz by allotting two points for every blank, one point for the correct word, and a second point for the correct spelling of the word. Let the students know beforehand how the quiz will be graded.

You may also wish to develop paragraphs that explain biodiversity and its importance, and extinction and its effects.

NOTES

NOTES

A Vote for Ecology

Activity 1 Diversity in Living Things

GOALS

In this activity you will:

- Observe a group of diverse organisms.
- Relate the structure of an organism to its adaptation to the environment.
- Describe the organization of the biosphere.
- Define biodiversity and explain its importance.
- Explain the effects of human activity on biodiversity.
- Read about the effects of extinction.
- Practice safe laboratory techniques for handling living organisms.

What Do You Think?

It is estimated that 4% of all living species are found in Costa Rica, even though this country comprises only 0.01% of the area of the Earth.

- **How many species do you think are found in Costa Rica? How many species are found globally?**
- **Why do you think that Costa Rica has such a large number of species?**

Write your answers to these questions in your *Active Biology* log. Be prepared to discuss your ideas with your small group and other members of your class.

For You To Do

This activity provides you with an opportunity to view several very different species of organisms. It should give you an appreciation of the huge diversity of life that fills your world.

What Do You Think?

- Costa Rica provides a habitat for over 500,000 species, including more than 800 species of ferns, 10,000 kinds of flowering plants, 850 bird species, 3000 butterfly species, and 209 species of mammals.

  Scientists estimate that more than 10 million different species inhabit the Earth. Some suggest that the number may be as high as 100 million. Yet, of the number of species that are estimated to exist, only about 1.75 million have been discovered and named.

- Costa Rica is in a tropical location, and therefore contains many more species than a country in a temperate or cold zone. It is a strip of land, relatively young in geological terms, that connects North and South America. As such, it provided a pathway for interchange between North and South America. Also, it has a varied topography of mountains, valleys, and lowlands, resulting in many microclimates, supporting diverse ecosystems.

Student Conceptions

It is doubtful that students will comprehend the incredible diversity of life on Earth. It is doubtful that many of us can! You may wish to begin by having each student contribute the name of one or two different species of organisms to form a class list. Encourage them to take into account all life forms including fungi and microbial organisms in addition to plants and animals.

Although students generally understand that ecology is the study of relationships, they often limit the relationships to those among living organisms. Encourage the students to realize that abiotic factors play a significant role in an ecosystem.

Part A: Observing Animal Diversity

1. With your teacher, review the guidelines concerning the proper handling of laboratory animals. Follow these guidelines carefully.

2. In your *Active Biology* log make an enlarged copy of the table shown below. The table should extend across two facing pages. Each of the 13 spaces should allow for several lines of writing.

3. In the *Characteristics* column, copy the words in italics from each of the following questions. The 13th space is for any other observations you make. All the specimens of one animal species and the materials and equipment needed for observing them are arranged at the station. Each team will have a turn at each station. Record only your observations, not what you have read or heard about the organism.

   1 What is the *habitat* of the animal? Does it live in water, on land, or both?

   2 Is *body symmetry* radial (symmetry about a center) or bilateral (the left and the right sides of the body are mirror images)?

   3 Does the animal have a *skeleton* (a structure that supports the organism's body)? If it does, is it an endoskeleton (on the inside) or an exoskeleton (on the outside)?

   4 Is the animal's body *segmented* (divided into sections) or is it *unsegmented*?

Several of the activities that follow involve the use of organisms in water. The water that the organisms are in should be considered a contaminant. Tables, equipment, and hands should be washed carefully so that germs are not inadvertently passed to people.

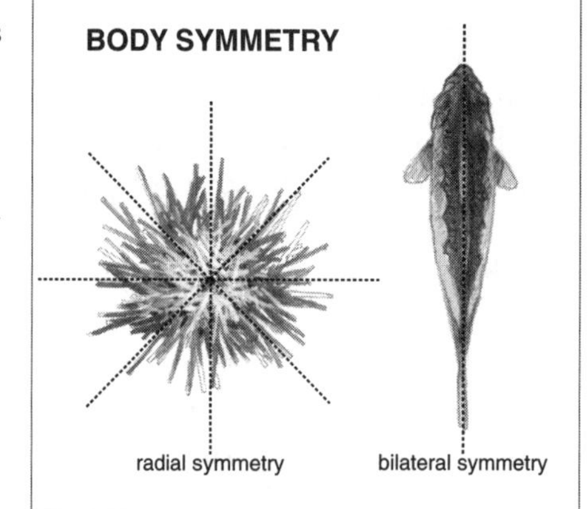

Comparing Animals						
	Characteristics	Hydra	Planarian	Earthworm	Hermit Crab	Frog
1						
2						
3						
13						

For You To Do

Teaching Suggestions and Sample Answers

Part A: Observing Animal Diversity

1. Establish with the students your guidelines for handling live animals.
The following are concerns you may wish to discuss with your students.
All animals must be treated as humanely as possible. Avoid bringing wild animals into the classroom. Obtain healthy animals from pet stores or biological supply houses. Earthworms may be an exception, but return any earthworms you collect to their natural habitat after your observations. Be sure that the proper environment can be established in the classroom for any live animal you use.
Under no circumstances inflict any pain or injury on the animal.
Avoid loud and startling noises.

   Review all the safety tips provided in the student text. Stress the importance of washing hands and avoiding any contact with the mouth and eyes. Frogs can harbor Salmonella. Hermit crabs can grasp tightly; warn students to keep their fingers away from the pincers.

Teaching Tip

To help students understand the different types of symmetry, ask them to classify a number of familiar items as having either radial or bilateral symmetry.
For example: what type of symmetry does each of the following show: a baseball bat (radial); a garbage pail (radial); a chair (bilateral) and a spoon (bilateral).

The following data table contains possible observations that the students may make, although the observations are presented in a more scientific language than you would expect of your students. The italicized sections provide additional information for the teacher. Encourage students to make their own observations and express them in their own words. Note in **Step 3**, students are advised to "Record only your own observations, not what you have read or heard." As you visit the stations, avoid the temptation to provide additional information, instead use suggestions that will develop the students' observational skills.

Blackline Master Ecology 1.1: Comparing Animals

Characteristics	Hydra	Planarian	Earthworm	Hermit Crab	Frog
1 Habitat	Water	Water	Land	Both	Both
2 Body symmetry	Radial Symmetry	Bilateral Symmetry	Bilateral Symmetry	Bilateral Symmetry	Bilateral Symmetry
3 Skeleton	No Skeleton	No Skeleton	No Skeleton	External Skeleton	Internal Skeleton
4 (Un)Segmented	Unsegmented	Unsegmented	Segmented	Segmented body, with some segments fused	Segmented backbone
5 Digestive cavity	Saclike gut with single opening	Saclike gut with single opening	Tube-like gut open at both ends (has a separate mouth and anus, the mouth is visible at the anterior end and the anus is visible at the posterior end)	Tube-like gut open at both ends (has a separate mouth and anus)	Tube-like gut open at both ends (has a separate mouth and anus)
6 Appendages	No appendages	No appendages	No appendages; the earthworm has bristles on its lower or ventral side, the dorsal side feels smooth, while the ventral side feels rough	Paired, jointed appendages	Paired, jointed appendages
7 Obtain oxygen	Oxygen diffuses into individual cells, no respiratory system present	Oxygen diffuses into individual cells, no respiratory system present	Oxygen diffuses from the air in the soil through the moist skin into the capillaries; oxygen is carried to the body cells in the blood	Oxygen diffuses from the water through the gill tissue (the gills are enclosed in the branchial chamber, which functions as a lung and is located on the sides of the thorax, above the crab's legs	Two nostrils are found on the tip of the head, permitting the frog to breath air
8 Sense organs	Sense organs are not visible, but must be present as the hydra responds to touch and the movement of the tentacles shows coordination (true nerve cells are found in hydra; they form a nerve network, which sends signals in all directions)	A concentration of nerve tissue in the head area resembles a primitive brain; eyespots are visible (although the eyespots cannot detect images, they are sensitive to light)	Sense organs are not visible externally; (a dorsal ganglion, a brain, is found at the anterior end; a ventral nerve runs the length of the earthworm with small ganglia giving rise to nerves in each segment)	A pair of (compound) eyes on long stalks and two pairs of antennae, one long and one short	Two eyes protrude from the head, (behind each eye is a tympanic membrane that picks up sound)

Comparing Animals (continued)

Characteristics	Hydra	Planarian	Earthworm	Hermit Crab	Frog
9 Move	Moves by contracting (flattening) and then attaches itself to a surface; they glide or somersault (The ectoderm cells contain contractile fibers, when these fibers contract the hydra moves)	Moves freely in a gliding motion	Moves by contracting body segments (there is a combination of circular muscles that surround the body, and longitudinal muscles that run the length of the body)	While walking, hermit crabs extend their antennae, claws, and two pairs of walking legs out of the shell opening.	Two large muscular hind legs permit the frog to jump great distances and the feet of the hind legs have five fingers that are webbed, allowing the frog to swim
10 Movement	The tentacles move to capture food	Able to change shape and direction	Body segments contract in a wave that runs along the length of the body	Antennae move and eyes move	Eye move, breathing movements can also be seen
11 Capture food	Tentacles stuff food into the mouth (hydra have specialized stinging cells used for defense and capturing food)	Sucks in small bits of food into the pharynx (the mouth is on the underside of the flatworm, a pharynx, which is a muscular tube can be extended out the mouth for feeding)	Soil enters the mouth as the earthworm burrows through it (the food is then sent backwards through the digestive system)	When feeding, hermit crabs extend their claws out of the shell opening to grasp their prey and put it into their mouth.	The frog flips out its tongue to capture its prey
12 Touched	Retracts and flattens in response to being touched	Moves away from the source of the touch	Moves away from the source of the touch (specialized receptors are sensitive to light, vibrations, chemicals, and heat)	Withdraws into its shells, blocking the entrance with its thick claws	Jumps away
13 Other observations	No digestive system	Digestive system consists of a mouth, a pharynx, and a highly branched intestine(simple excretory system consisting of a network of fine tubules opening to the outside via pores)	The earthworm has a digestive system that is a tube within a tube	Abdominal appendages are especially modified for keeping the shell firmly supported on the body	The two front limbs of the frog are short and have four fingers that are not webbed

Chapter 9

A Vote for Ecology

5 Which type of *digestive cavity* does the animal have, a sac (only one opening) or a tube (open at both ends)?

6 Does it have *paired appendages*? Are the limbs (arms, legs, fins, wings) found in pairs?

7 How does the animal *obtain oxygen*? Through lungs, gills, skin, or a combination of these?

8 Are any *sense organs* visible? If so, what types and where?

9 How does the animal *move* from one place to another?

10 Does it make any types of *movement* while it remains more or less in one spot?

11 How does the animal *capture* and take in *food*?

12 How does it react when *touched* lightly with a small brush?

Station 1: Observing Hydras

1. Place a single hydra in a small watch glass with some of the same water in which it has been living. Wait until the animal attaches itself to the dish and extends its tentacles. Then slowly add a few drops of a daphnia culture with a dropping pipette.
2. Touch the hydra gently with a soft brush. Observe its reactions.
3. Examine a prepared slide of a lengthwise (longitudinal) section of a hydra under a compound microscope. Try to determine the presence or absence of a skeleton and of a digestive system.

As you move among the stations, keep your hands away from your mouth and eyes. Wash your hands well after the activity.

Station 2: Observing Planarians

1. Place one or two planarians in a watch glass containing pond or aquarium water. Add a small piece of fresh raw liver. Observe using a stereomicroscope or hand lens.

2. Use a compound microscope to examine cross sections of a planarian. Examine whole mounts with a stereomicroscope. Determine the presence or absence of a skeleton and a digestive system.

Station 3: Observing Earthworms

1. Pick up a live earthworm and hold it gently between your thumb and forefinger. Observe its movements. Do any regions on the body surface feel rough? If so, examine them with a hand lens.
2. Place a worm on a damp paper towel. Watch it crawl until you determine its anterior (front) and posterior (back) ends. Use a hand lens to see how the ends differ. Describe.

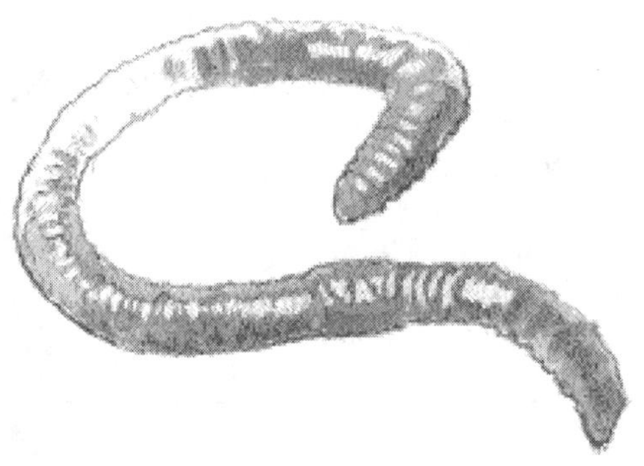

3. Place an earthworm on loose soil and observe its movements as it burrows.
4. Examine a model or a diagram of a cross section and lengthwise section of the earthworm's body.

Station 4: Observing Hermit Crabs

1. Observe the movements of the appendages and the pattern of locomotion (movement from one place to another) of a living land hermit crab. Observe the antennae. Touch them gently with a soft brush. Note the animal's reaction.

If you are observing a live crab in your classroom, keep your fingers away from the crab's pincers.

2. Place a small piece of food from the food dish in with the hermit crab. Observe how the hermit crab eats.

Station 5: Observing Frogs

1. Observe the breathing movements of a frog while it is not moving.
2. Observe the variety of movements of a live frog.
3. If possible, observe a frog capturing its food and feeding.

Wash your hands thoroughly before leaving the laboratory.

If you are handling a live frog in the classroom, do not rub your eyes. Wash your hands immediately after handling.

Chapter 9

Part B: Animal Adaptations to the Environment

1. Review what you have learned about each of the organisms in **Part A.** By reading across the table, you should be able to compare and contrast the characteristics of the five animals you have studied.

   a) For each animal, select two functions it performs as part of its way of life. Describe how its structure enables it to perform these functions.

Bio Talk

BIODIVERSITY

Organization in the Biosphere

The **biosphere** is the area on Earth where living organisms can be found. Most are found in a narrow band where the atmosphere meets the surface of the land and water. Life forms are referred to as the **biotic**, or living, component of the biosphere. The **abiotic**, or nonliving, component is made up of items like rocks, soil, minerals, and factors like temperature and weather.

Bio Words

biosphere: the area on Earth where living organisms can be found

biotic: the living components of an ecosystem

abiotic: the nonliving components of an ecosystem

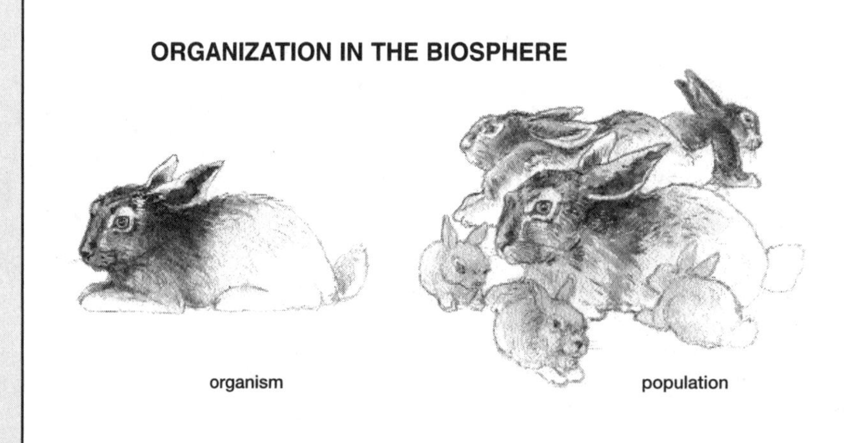

Part B: Animal Adaptations to the Environment

1. a) Student answers will vary. Some possible answers are provided.

Hydra

Feeding–tentacles are capable of working together to capture food

Gas exchange–hydra are only two *(germ)* layers *(ectoderm and endoderm)* thick, therefore each cell is in contact with water and oxygen is able to diffuse into each cell

Movement–hydra have nerve and muscle tissues that enable them to move

Planarian

Feeding–a digestive system is present for digesting food

Movement–a nervous system permits more complex movement and the planarian can move about freely

Gas exchange–the planarian is small and lives in the water where oxygen is free to diffuse into and out of each cell

Earthworm

Movement–the shape of the earthworm is ideal for burrowing through the soil; bristles along the underside of the earthworm help in its movement along and through the soil

Feeding–the "brain" and the mouth are at one end of the earthworm

Gas exchange–oxygen diffuses through the skin as earthworms burrow through moist soil

Hermit Crab

Movement–jointed legs enable the crab to move freely

Feeding–claws are able to grasp prey; eyes can see prey

Frog

Movement–strong muscular hind legs are used for jumping; feet of hind legs are webbed for swimming

Feeding–a long *(sticky)* tongue is used to capture prey

Response to danger–large eyes are able to move to see any "enemies" *(predators)*

Bio Talk

Teaching Tip

Students often think of an ecosystem in terms of the living or biotic components. They must appreciate the significance of the abiotic components as well.
Each ecosystem has a particular set of abiotic factors. These influence the biotic factors that exist in an ecosystem. Abiotic factors include: temperature, water, atmosphere and wind, fire, gravity, topography, and geological substrate and soil. Discuss with the students how each factor affects the biotic factors. For example, most species of organisms are limited in the range of temperatures in which they can survive. Water is one of the most important factors because it is essential to living things. The carbon dioxide required for photosynthesis and the oxygen required for respiration come from the atmosphere. Fire can be destructive, but it can also rejuvenate an aging forest. Many structural adaptations of plants and animal are related to overcoming the pull of gravity.

Just as you did in this activity, ecologists begin their studies with the **organism**. Their investigations are designed to explore how the individual interacts with its biotic and abiotic environment. However, an organism does not live on its own. It tends to form a group with others of the same **species**. (A species is a group of organisms that can reproduce successfully only with others of the same type.) These groups of species are called **populations**. When more than one population occupies an area, a **community** of organisms is created. The abiotic component as well as the community form a functional unit known as an **ecosystem**.

The Importance of Biodiversity and the Human Threat

In this activity you looked at some very different species of organisms. Scientists have discovered and named close to two million species. That would mean looking at a lot of different organisms. Yet, it may be less than 20 percent of the species that exist! There are thousands of organisms in the world that scientists know very little about. More than 750,000 species of insects have been identified. Yet, it is thought that at least twice that many exist. Biological diversity, or **biodiversity**, is the sum of all the different types of organisms living on Earth.

Bio Words

organism: an individual living thing

species: a group of organisms that can interbreed under natural conditions and produce fertile offspring

population: a group of organisms of the same species occupying a given area

community: all the populations of organisms occupying a given area

ecosystem: a community and the physical environment that it occupies

biodiversity: the sum of all the different types of organisms living on Earth

Teaching Tip

To help students understand the difference between an organism and a population and between an organism and a community you may wish to present the students with the following:

> In nearly any natural ecosystem that is not disturbed by human activities, only slight changes occur with the passage of time. Year after year, about the same number of frogs live in a woodland pond, and about the same number of flowers bloom nearby. The number and kinds of insects and soil microorganisms also remain about the same.
>
> The life of an individual organism is a striking contrast to this stable life. An individual exists as part of a population, and a population exists as part of a community. Often, the individual organism goes through a complex growth pattern during which its activities may vary tremendously. (You may wish to refer to the frog that the students observed in the **For You To Do** section.) It matures and produces young. Eventually the organism dies, but the community continues to exist. The stability of the natural community is the result of the interactions of individual organisms in the community with each other and with the environment.

The illustration on pages 536 and 537 of the student text is available as **Blackline Master Ecology 1.2: Organization in the Biosphere.**

A Vote for Ecology

Unfortunately, many organisms are disappearing. This is partly due to alterations of habitats. The result is a decrease in biodiversity. Ecosystems with a large number of different types of organisms are quite stable. Ecosystems with a small number of different organisms are less stable. Humans are partly responsible for this change. As the human population grows it occupies more land. This infringes on or destroys the habitats of many organisms.

Smog is also hazardous to people. This is especially true of those with respiratory problems, the elderly, and children. People have died from the effects of smog.

The smog created by automobiles and industry is killing many types of trees over a wide area of southern California. The needles of ponderosa pines, for example, gradually turn brown. The tops of palm trees have only small tufts. When this happens, photosynthesis is greatly reduced. The plants die. The Everglades National Park in southern Florida depends on a slowly moving sheet of water. The water flows from north to south. Drainage ditches built at the northern edge of the Everglades have decreased the flow of water over the entire area. As a result, many alligator holes have dried up. These holes helped to contain fires in the Everglades. Now, destructive fires are more frequent in this national park.

Tropical rainforests are the most diverse ecosystems on Earth. They are home for many different species. Two-thirds of the world's species are located in the tropics and subtropics. The cutting of trees in the rainforests today has grown at a rapid rate. The trees are cut for lumber, grazing land, and other uses. This loss of habitat is destroying many species every day. Nearly half of the Earth's species of plants, animals, and microorganisms will become extinct, be gone forever, or be severely threatened, during the next 25 years.

Bio Words

extinction: the permanent disappearance of a species from Earth

To find a similar rate of **extinction** (loss of species), you need to go back 65 million years. That was the end of the Cretaceous period when dinosaurs and other organisms disappeared. Because there are more

Teaching Tip

At this point students are introduced to the consideration of humankind and its relationship with the environment. Your students are part of a society that is going to have to continue to come to grips with such problems as air and water pollution, overpopulation, living space, and food production. As future voters they must have a certain understanding of these problems if they are to help set a healthy destiny for humankind in the future. Their **Chapter Challenge** to provide booklets for voters is actually an invitation to educate themselves about environmental issues that will affect their futures and their future votes.

species today than there were then, the absolute number of species lost will be greater now. Hundreds of species of plants and animals are threatened today. They include the whooping crane and some rare pitcher plants. Extinction is a natural process. However, the process has been speeded up because humans have changed whole ecosystems.

In tropical regions, humans are cutting down, burning, or otherwise damaging the rainforests. Extinction of many species as well as change in global climate are some of the effects of this deforestation.

Why is biodiversity important? Why does it matter if whooping cranes and pitcher plants become extinct? One argument comes from genetics. In a field of crop plants planted by humans, all the plants are genetically similar. They have all inherited the same characteristics.

About 90% of the world's food comes from 15 species of plants. Three of them are corn, wheat, and rice. However, there are over 10,000 known species of cereals.

Teaching Tip

Humans today rely very heavily on only a few species of plants for food. Tropical rainforests are being destroyed to grow food crops. What makes the matter even worse, is that the soil in tropical rainforests is not well suited for monocultures (growing a single species of plant to the exclusion of others). These soils need the renewal of decomposed material to keep an adequate level of nitrogen and phosphorus. A few seasons after planting, the soil will no longer support the growth of crops.

The greatest biodiversity exists in rainforests and many species are yet to be classified. All these organisms are being destroyed. However, students must also take into consideration the people who live in these areas. Many grow the cereal grains to feed their families and sustain an income.

A Vote for Ecology

If one individual gets a disease, all the plants may die. In a wild population a vast pool of genetic characteristics are available. This means that some of the plants could resist the disease. Therefore, not all the plants would be destroyed. The extinction of each wild population erases genetic material that could mean healthy crops and animals. Once extinction occurs, the genetic material is gone forever.

A second argument is related to the fact that simple ecosystems are unstable. Think of a field of corn as a simplified ecosystem. Suppose all the corn dies. This means that the whole ecosystem would collapse. The simpler the ecosystem, the easier it is to disrupt its balance. The fewer the species, the easier it is to upset an ecosystem. New species are evolving all the time. However, the process is very slow compared to the rate at which humans are able to cause species to become extinct. Each time a species becomes extinct, the biosphere is simplified a little more. It becomes more difficult to maintain the stable biosphere on which all life depends.

A third argument comes from research on plants. The island of Madagascar, off the east coast of Africa, is the only known habitat of the Madagascar periwinkle. This plant produces two chemicals not produced by other plants. Both of these chemicals are used to fight Hodgkin's disease, a leukemia-like disease. As the human population on Madagascar grew, the habitat for the periwinkle shrank. The periwinkle almost became extinct. Fortunately, botanists collected and grew some of these plants before they were gone forever. The medicines made from the Madagascar periwinkle are worth millions of dollars each year. They also help many people with Hodgkin's disease to live longer. These medicines never would have been known if the plant had become extinct.

Extinction Can Cause a "Domino Effect"

Every organism in an ecosystem is connected to all the other organisms. The reduction in biodiversity caused by the extinction of a single species can cause a "domino effect." The removal of one part from an ecosystem, like the removal of a moving part from a car, can cause the collapse of an entire food chain. If a species acts as a predator, it keeps the population of its prey in check. If a species is prey, it provides an important food source.

NOTES

For example, sea otters were over-hunted along the Pacific coasts of Asia and North America. This removed the main predator of the sea urchin. Predictably, the number of sea urchins grew rapidly. Sea urchins eat kelp, a form of seaweed. As the number of sea urchins grew, the amount of kelp declined. As a result, the fish that relied on the kelp for habitat and food were reduced in number.

Sea otters very nearly became extinct due to hunting pressure. For humans, killing the sea otters for their fur resulted in a decline in a valuable fishery. Where the sea otter has been reintroduced, sea urchin populations have fallen, kelp beds are being re-established, and the number of fish is increasing.

Restoring the Balance Is a Difficult Task

Introducing the sea otter to the Pacific northwest is an example of an attempt to restore a natural balance. It is not always easy to do. Conservationists have also tried to restore the whooping crane. In spring, whooping cranes fly north to live in the marshes and swamps of the prairies and the Canadian north. There they eat crayfish, fish, small mammals, insects, roots, and berries. Efforts by the United States and Canada have helped increase the population from a low of 14 individuals in 1940 to 183 in 1999. The whooping crane may be a success story, and it may not. Chemical pesticides were the original human threat to the crane. However, it was already struggling.

During the fur trade southern sea otters were hunted to near extinction. They are still a threatened species, and may very well be endangered.

NOTES

Once widespread throughout North America, the whooping crane wild population dipped to just 15 birds in 1937. Through conservation efforts the whooping crane has begun a slow recovery. However, coastal and marine pollution, illegal hunting, and the draining of wetlands continue to threaten the species.

Cranes must fly a long way between their summer homes in the north and their winter homes on the Gulf of Mexico. Along the way they are vulnerable to hunting and accidents. In addition, the whooping crane reproduces very slowly. Each year females produce two eggs, however, only one will mature. The first fledgling to crawl from the egg kills its brother or sister. This ensures there will be enough food for the survivor. However, it is very difficult for the species to increase its numbers.

Scientists do not understand all the relationships between species ecosystems. They cannot predict what will happen if biodiversity is reduced, even by one species. If one species becomes extinct, it could be disastrous. The extent of the disaster may not be known until later. Sometimes the balance cannot be restored.

Reflecting on the Activity and the Challenge

In this activity you observed several very different living organisms. You then discovered that there are millions of other different organisms alive on Earth. There are reasons why it is important to make sure that these organisms do not disappear forever from the Earth. For your **Chapter Challenge** you may choose to research an issue that relates to the disappearance of a given species. You can now explain why it is important to maintain biological diversity. Whether or not your issue deals with biodiversity, the public still needs to understand why biological diversity should concern them. You need to provide the meaning and importance of biodiversity.

Assessment Opportunity

You may provide your students with a quiz to assess how well they understood the reading. The following questions may be included:

1. What is the name for the area on Earth where living organisms can be found?

   (The biosphere)

2. Beginning with an individual organism, outline the organization of the biosphere.

   (Individual organism, population, community, ecosystem)

3. Explain what an abiotic factor of an ecosystem is and give three examples.

   (The nonliving components of an ecosystem; rocks and soil, water, temperature, atmosphere)

4. What is biodiversity?

   (The sum of all the different species of organisms living on Earth)

5. Define extinction.

   (The permanent loss of a species from Earth)

Biology to Go

1. Choose and identify two very different ecosystems.
   a) For each, name some of the populations that might be found in each community.
   b) Describe some of the abiotic factors that could affect each population.
2. What is biodiversity?
3. Explain how humans can influence biodiversity by changing the environment.
4. Why is maintaining biodiversity important?
5. a) Give an example of an ecosystem that has a high biodiversity.
   b) Give an example of an ecosystem that has a low biodiversity.
6. Choose an organism other than one that you studied in this activity. List at least three structures that have helped the organism adapt to its environment. Describe how each helps the organism live in its ecosystem.

Inquiring Further

1. The passenger pigeon and the human influence

Just over a century ago, the passenger pigeon was the most numerous species of bird on Earth. In the Eastern United States they numbered in the billions, more than all other species of North American birds combined. On September 1, 1914, at 1:00 PM the last surviving passenger pigeon died at the age of 29. Research and report on how humans were involved in the extinction of the passenger pigeon.

2. Extinction is forever

Humans were directly responsible for the extinction of passenger pigeons. However, this bird is not the only organism that has been threatened by humans. Research and report on another organism whose existence has been or is endangered by humans.

White rhinos are so large and powerful that in nature they must give way only to the elephant. Yet, humans are a major threat to their existence.

Biology to Go

1. Student answers will vary.
   a) Coral reef–living coral, algae, fish, sea urchins, crabs, sharks.
   b) The temperature of the water would be a very important abiotic factor as well as the wind and other weather conditions.

   a) Rain forest–palms, tree seedlings, vines and other climbing plants, tall trees, insects, birds, gorillas and monkeys.
   b) Temperature, rainfall, and soil are important abiotic factors in a rainforest.

2. Biodiversity is the sum of all the different types of organisms living on the Earth.

3. Humans destroy natural habitats as their population increases and they need to occupy more land. Smog from automobiles and industry can destroy trees. Land that is used for agriculture also infringes on natural habitats and can reduce a biologically diverse ecosystem into a monoculture.

4. (The reading presents three arguments; students may add their own reasons.)
   – If plants and animals are genetically similar a disease could destroy an entire population.
   – Simple ecosystems are unstable. Therefore, if one component is removed, the entire ecosystem could collapse.
   – Many plants are yet to be identified. Some of them may be beneficial to humans.

5. Student answers will vary.
   a) Rain forests have a very high biodiversity.
   b) The Arctic tundra would have a low biodiversity.

6. Student answers will vary. Students may choose an animal that lives in a cold climate and indicate that thick fur, fat layers, and hibernation are adaptations to this environment. In a hot climate students might indicate that animals are active at night to avoid the heat of day, and plants have spines instead of leaves to conserve water.

Inquiring Further

1. **The passenger pigeon and the human influence**
   "When an individual is seen gliding through the woods and close to the observer, it passes like a thought, and on trying to see it again, the eye searches in vain; the bird is gone." John Audubon on the passenger pigeon. The passenger pigeon is an example of probably the most terrible mass slaughter in the history of wildlife. Students should not have any difficulty finding information on this subject.

2. **Extinction is forever**
   Students may wish to research either an extinct animal or one that is endangered. Animals, other than the passenger pigeon, that are extinct as a result of humans include the quagga (a horse-like relative of the zebra), the great northern sea cow, the spectacled cormorant, and the dodo bird.

Blackline Master Ecology 1.2: Organization in the Biosphere

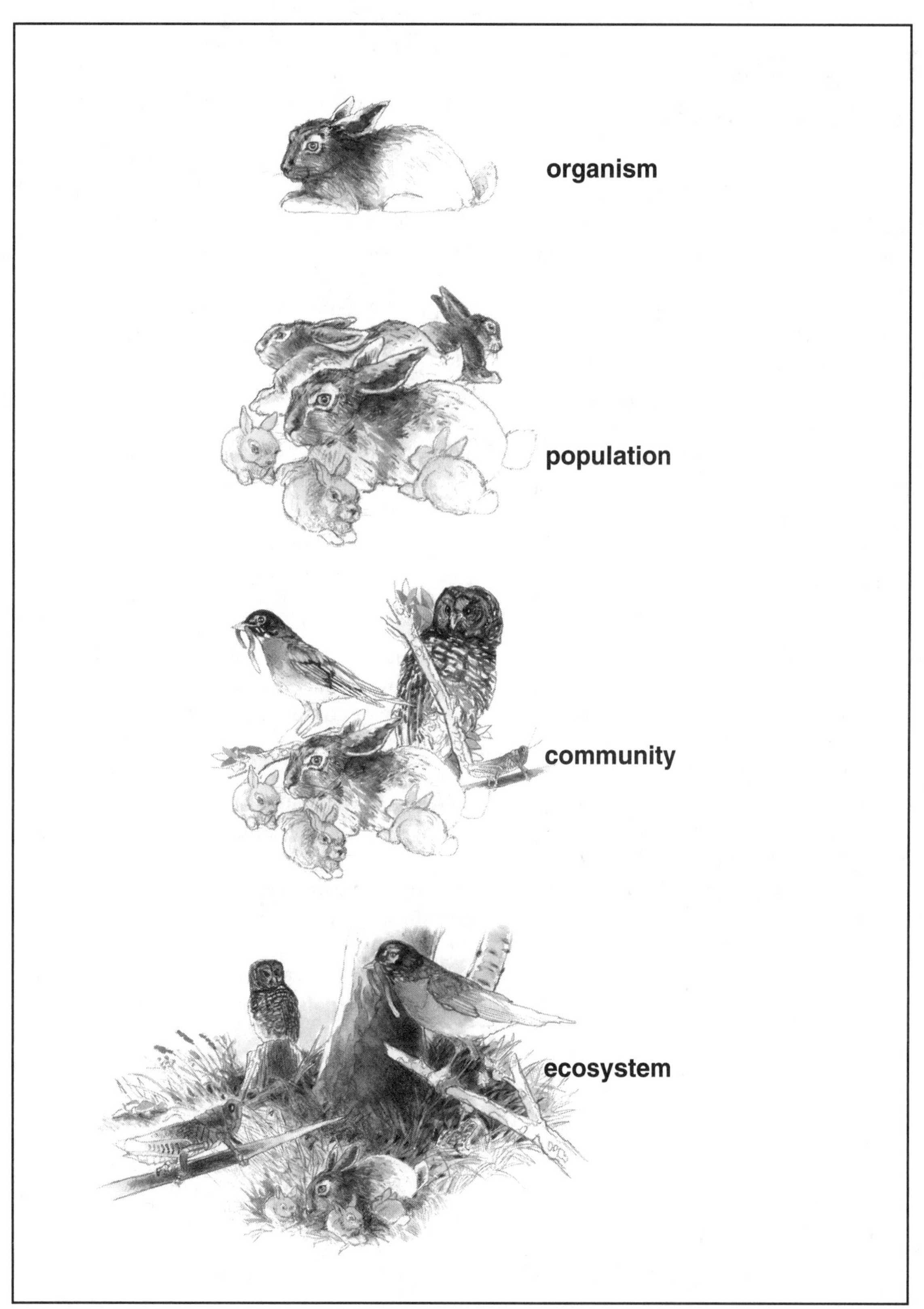

ACTIVITY 2– WHO EATS WHOM?

Background Information

Food Chains and Food Webs

Most food webs can be misleading in the sense that they emphasize the herbivore and the carnivore species. For example, in some forest ecosystems, only 1% of the producer biomass is consumed by the herbivores. The additional 99% ends up as material that enters the detritus food web. The **BioTalk** presents to the students the vocabulary they need to explain both a food chain and a food web. You may wish to further emphasize the importance of the decomposers.

The Decomposer Food Chain

A soil ecosystem is like a city in that its food supply comes from the outside. Most soil organisms are consumers and live below the surface of the soil. In soil, food made during photosynthesis comes down into plant roots from the leaves and stems. Plant roots can serve as food for some consumers, but for most of them, the food supply comes from the remains of other organisms. Therefore, decomposers like fungi and bacteria are very important in soil communities.

A dead root in the soil still contains a large amount of starch; likewise, a dead animal still contains fats and proteins. Beetles and other small animals living on or just below the soil surface use these substances as food. Other substances, such as cellulose in plant bodies and chitin in insect bodies, can be broken down only by microorganisms and fungi. Decomposers that use cellulose and chitin as food leave simple organic substances as waste products. The wastes still contain energy that other fungi and bacteria can use. These fungi and bacteria, in turn, leave simple inorganic substances in the soil, which still other decomposers can use. Thus, one decomposer depends on another for its food supply.

This type of food chain is like an assembly line in reverse. Instead of building up step-by-step from simple to complex compounds, the decomposer food chain breaks down complex organic substances. At the end of the chain, only inorganic substances such as carbon dioxide, nitrates, ammonia, and water remain. Plants then can assemble these inorganic materials into organic compounds and begin the cycle again.

Some soil organisms produce substances that harm other organisms. These substances may accumulate in the soil around the organism that formed them, reducing the growth of competing organisms. We use these substances as antibiotics to combat bacterial infections. Auromycin, for example, is derived from a bacterium, and penicillin is derived from a fungus.

Goals and Assessments

Goal	Location in Activity	Assessment Opportunity
Distinguish between a food chain and a food web.	**For You To Do** Parts A and B **BioTalk**	**Biology to Go** Questions 3 and 6.
Explain the roles of the producers, consumers, and decomposers.	**For You To Do** Part A **BioTalk**	Students identify roles in a food chain that they construct; **Biology to Go** Questions 2 and 6.
Understand the meanings of autotroph, heterotroph, herbivore, carnivore, and omnivore.	**For You To Do** Part A **BioTalk**	**Biology to Go** Question 5.
Recognize the dependence of organisms on one another in an ecosystem.	**For You To Do** Part B **BioTalk**	**Biology to Go** Questions 1 and 7.

Activity Overview

This activity introduces students to food chains and food webs. Although most students will have studied these in earlier grades, it is worthwhile to take the time to reintroduce the students to the concept that energy flows from producers to consumers in a food chain. This will prepare them for the next activity in which they investigate energy flow in ecosystems. The vocabulary required to discuss the food web is also reintroduced. By forming a food web with string, it is hoped that students will gain a sense of how dependent one organism is on another.

Preparation and Materials Needed

Preparation

No preparation is required for **Part A.** However, the diagrams on page 545 are provided on **Blackline Master Ecology 2.1: A Food Chain.** You may wish to reproduce these for your students to cut out and assemble into a food chain.

For **Part B: A Food Web** you will need to prepare a name tag for each student. You will also need to provide each student with an organism card that indicates what the organism eats and what it is eaten by.
See **Blackline Master Ecology 2.2: Organism Cards.**

Materials/Equipment Needed

- ball of string about 35 m long (one for class)
- plastic name tag holders (one per student)

Learning Strategies for Students with Limited English Proficiency

1. Point out new vocabulary in context. Practice using the words as much as possible.

food chain	food web	autotroph	heterotroph
producer	consumer	herbivore	carnivore
omnivore	decomposer		

   Provide students with **Blackline Master Ecology 2.3: Vocabulary Practice.** Have students work in pairs to complete this sheet.

2. Provide students with a copy of **Blackline Master Ecology 2.1: A Food Chain.** Have them cut the diagrams apart and rearrange them to form a food chain. Ask the students to label the food chain using words in their first language and corresponding words in English.

3. Have the students work in pairs. Provide them with 20 (or more) index cards. Ask them to cut the cards in half. Have them write all the new vocabulary words on one half of a card. Tell them to write the definition of each word on the other half of the card. Explain that they will now play a game of "Go Fish" with the deck of cards. Shuffle the cards and deal each player 5 cards. One player begins by asking another player if they have the definition of "food chain," for example. If the second player has the card, he or she passes it to the first player. The first player lays the pair face down on the table. If the second player does not have the definition, he or she replies, "go fish." The student then picks up a card from the remaining cards.

   Continue to build on the deck of cards as new vocabulary is introduced.

A Vote for Ecology

Activity 2 Who Eats Whom?

GOALS

In this activity you will:

- Distinguish between a food chain and a food web.
- Explain the roles of the producers, consumers, and decomposers.
- Understand the meanings of autotroph, heterotroph, herbivore, carnivore, and omnivore.
- Recognize the dependence of organisms on one another in an ecosystem.

What Do You Think?

You have probably heard the question, "If a tree falls in a forest and there is no one there to hear it, does it make a noise?"

- **In ecology you might ask the question, "If a tree falls in a forest and there is no one there to haul it away, what happens to it?"**

Write your answer to this question in your *Active Biology* log. Be prepared to discuss your ideas with your small group and other members of your class.

For You To Do

In this activity you will have an opportunity to explore how organisms in an ecosystem are dependent on one another.

What Do You Think?

- Dead trees, whether still standing or fallen, create a lively ecosystem. Bacteria and fungi begin to decompose the tree and they in turn become food for other animals. Trees that fall into water provide resting places for frogs and birds and safe havens for fish. On land, the decomposers turn the tree into nutrient-rich soil in which ferns and other forest plants can live. It is not uncommon to walk through a forest and see another tree rooted in a decaying, fallen tree.

Student Conceptions

Some students may underestimate the role of decomposers in an ecosystem. Decomposers provide the link between the abiotic components and the biotic components of the biosphere. Without decomposers most of the organic matter would remain "tied up" in dead organisms.

Stress to the students the difference between the organisms that eat dead matter and garbage, the scavengers, and those that break down waste and dead materials, the decomposers.

Part A: A Food Chain

1. Look at the organisms in the pictures on the right.
   a) Link the names of the organisms together by the words "is eaten by."
   b) Show the relationship between the organisms as a linked word diagram or chain.
   c) Use arrows to show the direction in which food energy moves in the food chain you constructed in **Part (b)**.
   d) Identify the producer in the food chain.
   e) Identify the consumers in the food chain.
   f) Which consumer is a herbivore (feeds on plants)?
   g) Which consumers are carnivores (feed on other animals)?
   h) What elements are missing from this food chain?

Part B: A Food Web

1. Your teacher will provide you with a card that names an organism, what it does, what it eats, and what it is eaten by. You will also be given a name tag with your organism's name on it.

   Read your card and attach your "name" tag where others can readily identify you.
2. Clear a large area in your classroom, or find another large open area in or near your school. Form a large circle.

For You To Do

Teaching Suggestions and Sample Answers

Part A: A Food Chain

1. a) A green plant is eaten by a grasshopper, which is eaten by a frog, which is eaten by a snake, which is eaten by a hawk.
   b) and c) green plant–grasshopper–frog–snake–hawk.
   d) The producer is the green plant.
   e) The consumers are the grasshopper, frog, snake, and hawk.
   f) The grasshopper is a herbivore.
   g) The frog, snake, and hawk are carnivores.
   h) The decomposers are missing from the food chain.

Teaching Tip

Students who require extra individual attention may wish to use a copy of **Blackline Master Ecology 2.1: A Food Chain** to cut up and rearrange the diagrams on a sheet of paper. Ask the students to label each organism as a producer or consumer, and as a herbivore and a carnivore. Remind them that the arrows they draw should point in the direction of energy transfer.

Part B: A Food Web

1. Provide each student with a name tag and an organism card. See **Blackline Master Ecology 2.2: Organism Cards**. Alternatively, you may wish students to generate their own organism cards. If time permits, you may wish to develop organism cards that include pictures of the organism.

A Vote for Ecology

3. Obtain a big ball of string, about 35 m in length. Give the ball of string to one of the students.
4. The first student will say what organism he/she represents. Also, the student will indicate what the organism eats and what it is eaten by. The ball of string is then directed to one of students who represents the predator or the prey.
   a) As the game progresses, what appears to be forming in the center of the circle?
5. Suppose one organism is removed from the circle. Your teacher will direct you which organism will be removed.
   a) What happens to the web that was created?
   b) How does the removal of an organism impact on the other organisms in the circle?
6. Suppose that your circle has only a few organisms.
   a) What would happen to the web in this case if one of the organisms were removed?
   b) In which situation, a large or small "circle" of organisms, does the removal of an organism have a greater impact?

BioTalk

Bio Words

food chain: a series of organisms through which food energy is passed in an ecosystem

food web: a complex relationship formed by interconnecting food chains in an ecosystem representing the transfer of energy through different levels

autotroph: an organism that is capable of obtaining its energy (food) directly from the physical environment

heterotroph: an organism that must obtain its energy from autotrophs or other heterotrophs

producer: an organism that is capable of making its own food

consumer: a heterotrophic organism

herbivore: a heterotroph that feeds exclusively on plant materials

Food Chains and Webs

A bat ate a mosquito that had bitten a coyote that had eaten a grasshopper that had chewed a leaf. All these living things make up a **food chain.** A food chain is a step-by-step sequence that links together organisms that feed on each other. The story, however, is incomplete. It does not mention that many animals other than coyotes eat grasshoppers and mosquitoes bite other animals. It also does not consider that coyotes and bats eat and are eaten by a great many other living things. When you consider that the kind of plant a grasshopper might eat may also be eaten by various other consumers, you start to build a picture that links together a whole community of living things. Those links resemble a **food web** rather than a food chain. A food web is a series of interconnected food chains or feeding relationships. The diagram shows how members of a community interact in a food web.

Organisms in the Food Web

Autotrophs are organisms that are capable of obtaining their energy (food) directly from the environment. Most autotrophs obtain their energy through the process of photosynthesis. In this process solar

3. Yarn also works well in this activity.

4. a) The string will begin to form a web.

5. a) There will be slight "loosening" of the web.
   b) Ask the students to focus on what happens to the other organisms in the circle. Pay attention to the organisms that will now not be eaten as well as those that were going to eat the eliminated organism. Ask students to also consider if eliminating an organism affects organisms that are not directly linked.

   Students should note that eliminating one organism affects all the organisms in a food web. However, in a food web with a large number of organisms, there are alternative pathways possible.

6. You may wish to demonstrate this type of food web by using, for example, a tundra food web. Producers could include flowering plants, grasses, sedges, willows, and lichens. First-order consumers could include musk ox and lemmings. Second-order consumers could include snowy owls and the Arctic fox, both of which feed on lemmings.
   a) The web could completely collapse if an organism is removed.
   b) The greater the biodiversity of an ecosystem, the less the impact of removing one organism.

BioTalk

Teaching Tip

You may wish to provide students with **Blackline Master Ecology 2.3: Vocabulary Practice** to help in their vocabulary building.

energy is converted into a form of energy that can be used by the organism. **Heterotrophs** obtain their energy from autotrophs or other heterotrophs. For this reason autotrophs, the organisms that "make" the food, are called **producers**. In the diagram, grass, vegetables, and trees represent the producers. The heterotrophs are called **consumers**. **Herbivores** are first-order consumers. They feed directly on the plants. These organisms are removed by just one step

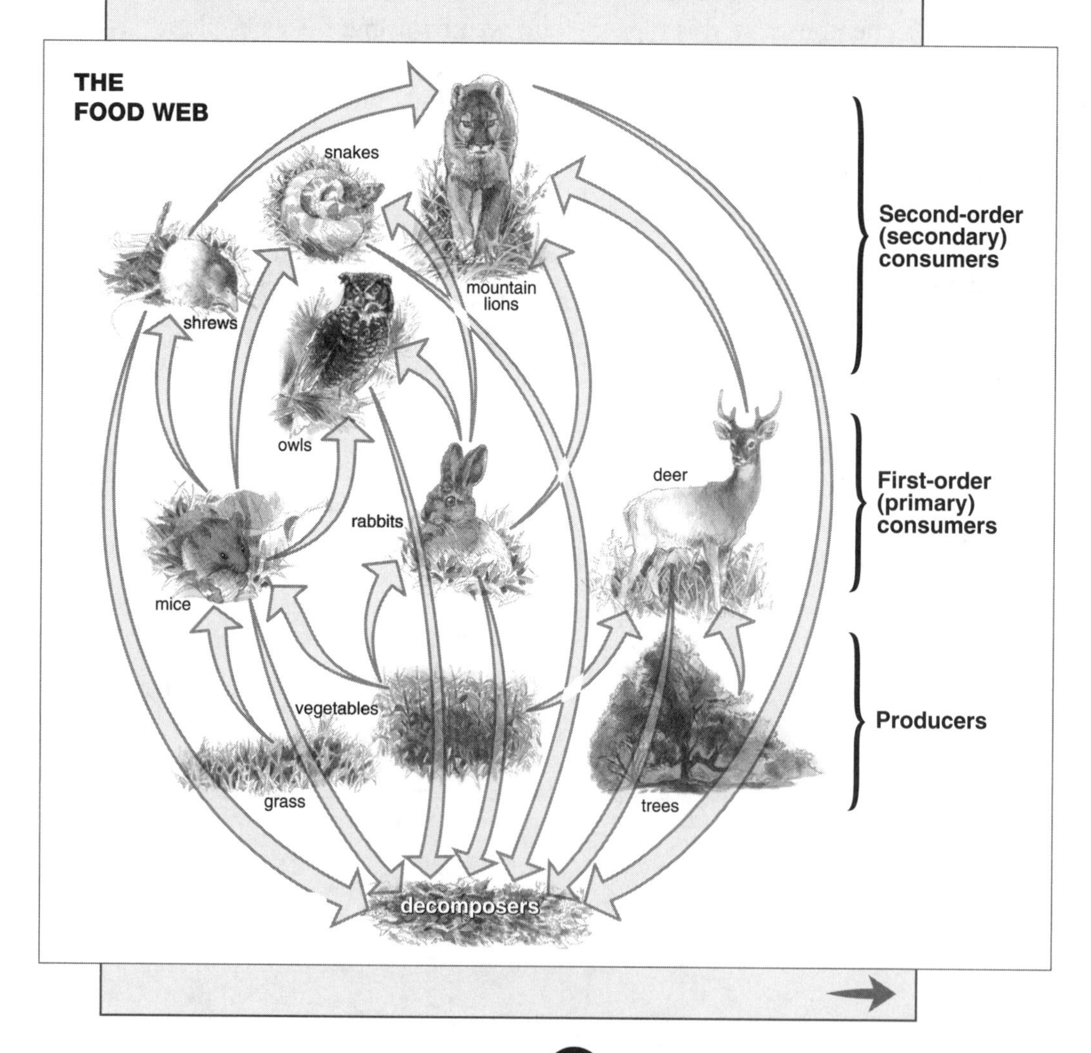

Teaching Tip

The diagram on student page 547 is provided as **Blackline Master Ecology 2.4: The Food Web.**

Emphasize to the students that the arrows point in the direction of the energy flow. Some students get confused and assume that the arrows should point from the predator to its prey.

A Vote for Ecology

Bio Words

carnivore: an animal that feeds exclusively on other animals

omnivore: a heterotroph that feeds on both plant materials and animals

decomposers: organisms that break down the remains or wastes of other organisms to obtain their nutrients

in the food chain from the producers. In this example, they include mice, rabbits, and deer. **Carnivores** are second-order consumers. They feed on the animals that eat other plants. The owl and the mountain lion are just two examples of carnivores. **Omnivores** eat both plants and animals. A human is an example of an omnivore.

There is another group of organisms in the food web that is so important that these organisms are often treated as a separate group. They are the **decomposers.** They break down the complex organic molecules that are found in the wastes and bodies of other organisms. They do this to obtain food energy for their own use. In the process, they release nutrients back into the ecosystem. Bacteria and fungi make up most of the decomposers.

Alternative Pathways Maintain Stability in Food Webs

The alternative pathways in a food web help maintain the stability of the living community. If the rabbits in some area decrease in number, perhaps because of some disease, the owls might be expected to go hungry. However, this is not the case. The rabbits eat less vegetation. Hence, the greater number of plants produces more fruits, and seeds and furnishes better hiding places for mice. Soon a larger population of mice is present. The owls transfer their attention from rabbits to

The food habits of rabbits vary depending on location, time of year, and species of rabbit. They generally prefer to eat tender, green vegetation. They also eat leaves, bark, seeds, and even fruit of woody plants. Rabbits begin feeding in the evening and continue throughout the night.

Teaching Tip

Emphasize once again the role of the decomposers. This will become more evident to the students when they study the cycling of matter later in this chapter.

mice. This reduces the danger for surviving rabbits, and these primary consumers have a better chance to rebuild their numbers. The greater the number of alternative pathways a food web has, the more stable is the community of living things which make up the web.

Owls are nighttime (nocturnal) birds of prey. Owls feed entirely on living animals. They eat everything from insects to mammals as large as rabbits. The size of the prey is proportional to the size of the owl.

Only a few of the possible offspring of a plant or animal survive to reproduce. Of all the seeds a plant forms, all but a few are eaten by animals. Some die from diseases. Others are killed by poor weather conditions. This can happen either as seeds or somewhat later in life, as young plants that have not yet formed seeds of their own.

Teaching Tip

Continue with the idea of animal adaptations to the environment introduced in the first activity. Have students look at the eyes of a herbivore and the eyes of a carnivore. The photographs on pages 548 and 549 are good examples to use. You can use the following questions to guide their observations:

- Where are the eyes of a herbivore located and in which direction are they facing?
(The eyes are located at the side of the head and are facing outward.)

- Where are the eyes of a carnivore located and in which direction are they facing?
(The eyes are located at the front of the head and face forward.)

- What advantage is the placement of the eyes to each type of animal?
(The placement of the eyes of a herbivore allow the animal to watch for approaching predators. The placement of the eyes of a carnivore allows the animal to watch its prey as it is chasing it and provides accuracy as it pounces.)

- Where are the eyes of a herbivore placed in relation to its mouth? What is the advantage of this adaptation?
(The eyes of a herbivore are often located well above and away from the mouth. This permits the animal to see approaching predators while feeding.)

- Where are the eyes of a carnivore placed in relation to its mouth? What is the advantage of this adaptation?
(The eyes of a carnivore are usually located close to its mouth. The mouth (or beak) is often used in killing the prey.)

A Vote for Ecology

Humans are so used to thinking of the welfare of their own species, that they tend to regard as "wasted" all the offspring that do not survive. But there is another side to the picture. For one thing, the world lacks space for so many individuals of any one kind. Also, these individuals are needed as food by a great variety of consumers. Without the fruits, seeds, young plants, and foliage, the primary consumers could not exist. Without the primary consumers, the plants would die. They would become overcrowded or lack nutrients. Without the primary consumers, the secondary consumers would be reduced in numbers because of competition, or would become extinct. Without waste from plants and animals, including dead remains, the decomposers would not be able to get their nutrients. Without decomposers, nutrients that the producers require would not be returned to the soil or water. Through the presence of all these components in the food web, each species is held in check, and the community maintains its stability.

Reflecting on the Activity and the Challenge

In this activity you looked at how every organism is dependent on other organisms and how they are all held together by a food web. You can now begin to understand how the stability of any ecosystem depends on each one of its components. In the **Bio Talk** reading section you were also reintroduced to many terms that are used by ecologists. In discussing your environmental issue, you will be expected to use these terms correctly. You will probably also want to explain the importance of some of these terms in your booklet to educate the public.

How are primary consumers a benefit to plants?

Teaching Tip

Ask the students to consider the question posed under the photograph:
How are primary consumers a benefit to plants?
The answer is provided in the reading above. Without primary consumers, plants might die from overcrowding and lack of nutrients.

Students may also suggest that some primary consumers help to pollinate plants. Also, burrs stuck to the fur of primary consumers help in the dispersal of seeds.

Biology to Go

1. In what ways are living organisms affected by other living organisms?
2. What is the role of decomposers in a biological community?
3. What is the difference between a food chain and food web? Use an example to explain your answer.
4. a) Why are autotrophs called the producers in an ecosystem?
   b) Why are heterotrophs called consumers?
5. Are you a herbivore, carnivore, or omnivore? Explain your answer to show that you understand the meaning of each term.
6. Create a food web that includes you and at least five other organisms. Identify the decomposers, producers, and consumers as you diagram your food web.
7. In which ecosystem would the removal of an organism disrupt stability more, an Arctic ecosystem or a deciduous forest? Explain your answer.

Water makes up the largest part of the biosphere. Aquatic regions, both freshwater and marine, are home to many species of plants and animals. As you inquire further into aquatic food webs, you may be surprised at how many different types of aquatic ecosystems exist.

Inquiring Further

Aquatic food webs

Water covers over two-thirds of the surface of the Earth. Research and construct an aquatic food web. Identify the producers and consumers.

Biology to Go

1. Living organisms are affected by one another in competing for food, for light, for types of soil, for water, for mates, for places to live, and in many other ways. They may affect each other adversely or may interact in ways that are helpful to one another. When one organism in a community is affected, it directly or indirectly affects all others.

2. The decomposers in a biological community return the chemical compounds of the bodies of dead organisms, and of the waste products of living organisms to the environment in a form usable by other organisms.

3. A food chain is a straight-line feed path in an ecosystem. For example, a plant is eaten by a rabbit, which is eaten by an owl. A food web links together a whole community of living things in a complex relationship of feeding paths. For example, a plant is eaten by a rabbit, deer, mice, caterpillars. Owls eat mice, rabbits, and caterpillars.

4. a) Autotrophs are able to obtain their energy directly from the physical environment. They can produce their own food, hence they are called producers.
   b) Heterotrophs must obtain their energy from autotrophs or other heterotrophs. Therefore, they consume other organisms to survive.

5. Students will answer that they are either omnivores or herbivores. Their answers must indicate that they understand that omnivores eat both plants and animals, whereas herbivores eat only plant material.

6. Student answers will vary. Suggest that students who are vegetarians nonetheless include some second-order consumers in their food webs.

7. The removal of one organism would have the greatest effect on an Arctic ecosystem. There is much less biodiversity in the Arctic, and therefore food webs in this region are not as stable. A deciduous forest has considerable biodiversity.

Inquiring Further

Aquatic Food Webs

Students will have no shortage of food webs to construct. Perhaps you could guide different students to research different types of aquatic ecosystems: rivers and streams, lakes and ponds, wetlands, shorelines, temperate oceans, tropical oceans, brackish water.

Blackline Master Ecology 2.1: A Food Chain

Blackline Master Ecology 2.2: Organism Cards

Organism: Seeds

Eats:

Eaten by: Mice

Organism: Vegetation (sword fern, dogwood, mountain laurel)

Eats:

Eaten by: Insects

Organism: Berries

Eats:

Eaten by: Raccoons

Blackline Master Ecology 2.2: Organism Cards

Organism:	Mice
Eats:	Seeds
Eaten by:	Owls, eagles, hawks

Organism:	Insects
Eats:	Vegetation
Eaten by:	Grouse, Townsend's Mole, Ring-necked Pheasants, Woodpeckers

Organism:	Raccoons
Eats:	Berries
Eaten by:	

Blackline Master Ecology 2.2: Organism Cards

Organism:	Owls
Eats:	Mice, Townsend's Mole
Eaten by:	

Organism:	Eagle
Eats:	Mice, Townsend's Mole
Eaten by:	

Organism:	Hawks
Eats:	Mice, Townsend's Mole
Eaten by:	

Blackline Master Ecology 2.2: Organism Cards

Organism:	**Grouse**
Eats:	**Insects**
Eaten by:	

Organism:	**Townsend's Mole**
Eats:	**Insects**
Eaten by:	**Owls, Eagles, Hawks**

Organism:	**Ring-necked Pheasants, Woodpeckers**
Eats:	**Insects**
Eaten by:	

Blackline Master Ecology 2.2: Organism Cards

Organism:	Plankton
Eats:	
Eaten by:	Krill, Small fish and squid

Organism:	Krill
Eats:	Plankton
Eaten by:	Adelie Penguins, Emperor Penguins, Crab eater seals, leopard seals, baleen whales

Organism:	Small fish and squid
Eats:	Plankton
Eaten by:	Wendell and Ross seals, large fish, toothed whales, Emperor Penguins

Blackline Master Ecology 2.2: Organism Cards

Organism:	Baleen Whale
Eats:	Krill, Plankton
Eaten by:	

Organism:	Adelie Penguins
Eats:	Krill
Eaten by:	Leopard Seals

Organism:	Emperor Penguins
Eats:	Krill, Small fish and squid
Eaten by:	Leopard Seals

Blackline Master Ecology 2.2: Organism Cards

Organism:	Crab-eater Seals
Eats:	Krill
Eaten by:	Leopard seals

Organism:	Leopard Seals
Eats:	Krill, crab-eater seals, Adelie Penguins, Emperor Penguins
Eaten by:	Toothed whales

Organism:	Wendell and Ross Seals
Eats:	Small fish and squid
Eaten by:	Toothed whales

Blackline Master Ecology 2.2: Organism Cards

Organism:	**Large fish**
Eats:	**Small fish and squid**
Eaten by:	**Toothed whales**

Organism:	**Toothed Whales**
Eats:	**Leopard seals, large fish, Wendell and Ross seals, small fish and squid**
Eaten by:	

Blackline Master Ecology 2.3: Vocabulary Practice

Name:______________________________________Date:____________________

Use the words at the bottom of the page. Place the letter for the word that best fits with each sentence below. You may use words more than once and some sentences may fit with more than one word.

1.______This is the word for a simple, straight-line feeding path.

2.______This organism is able to make its own food.

3.______This connects each organism to all the organisms it eats, and the organisms that eat it.

4.______This organism obtains its food only from plants.

5.______These organisms break down the wastes or remains of other organisms to obtain their food.

6.______This organism only eats other animals to obtain its food.

7.______This organism eats both plants and animals to obtain its food.

8.______This organism is able to get its energy directly from the physical environment.

9.______This organism must obtain its energy from other organisms.

10.______This organism is heterotrophic.

11.______A wolf is an example of this type of organism.

12.______This word can be used to describe the links between grain, a mouse, and a snake.

13.______A flower is an example of this type of organism.

14.______A rabbit is an example of this type of organism.

15.______A bear eats berries and fish and is an example of this type of organism.

16.______This word can be used to describe a grasshopper, a butterfly, a snake, a hawk, and even you.

A. autotrophy	B. carnivore	C. consumer	D. decomposer	E. food chain
F. food web	G. herbivore	H. heterotroph	I. omnivore	J. producer

Chapter 9

Blackline Master Ecology 2.4: The Food Web

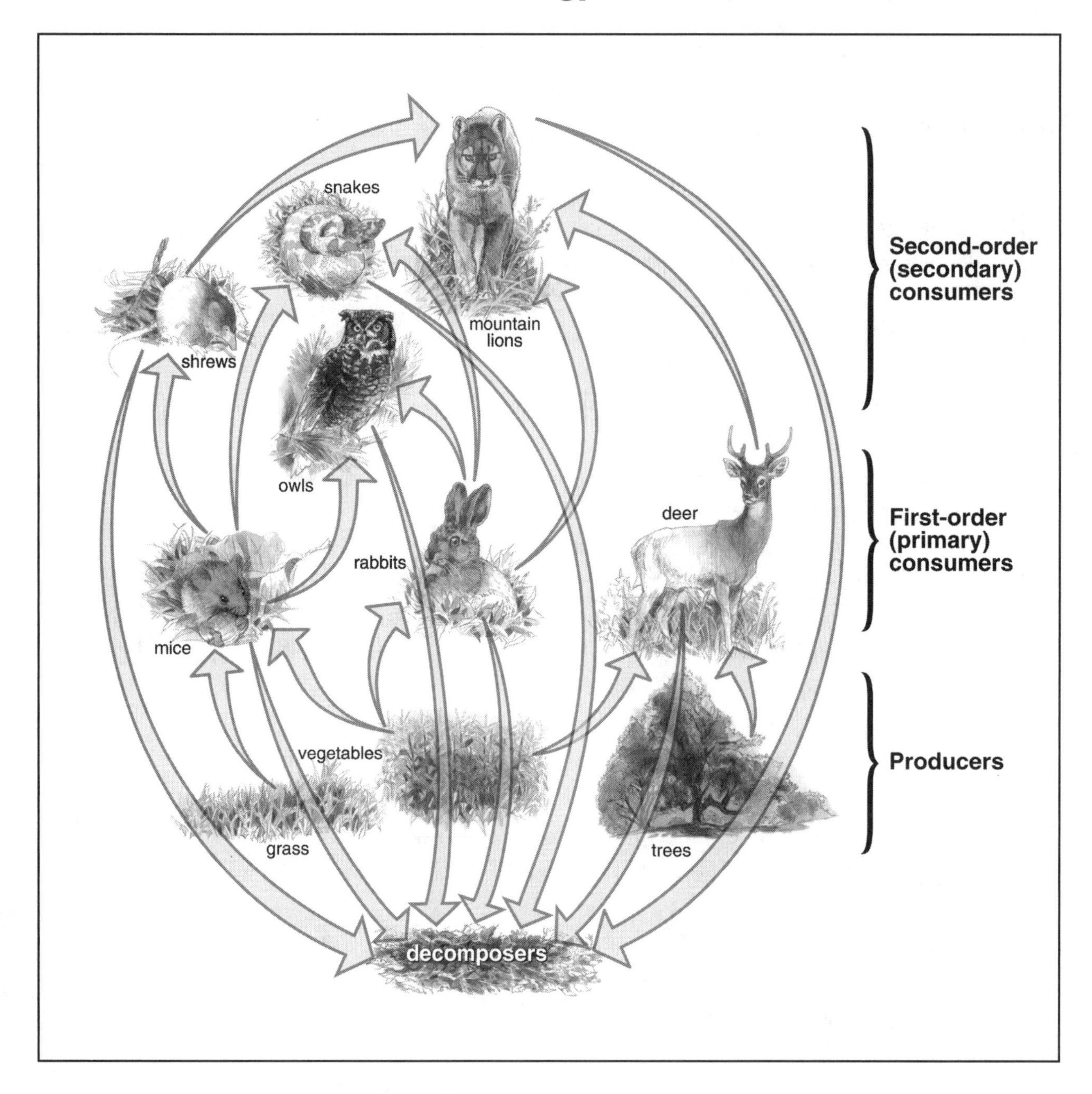

NOTES

ACTIVITY 3–ENERGY FLOW IN ECOSYSTEMS

Background Information

What is Energy?

The concept and importance of energy cannot be easily summed up in one statement. Loosely speaking, energy has to do with the motions of matter, or the potential motions of matter. Energy is often defined as the ability to do work. When work is done, energy is transferred from one object to another. Energy can take on many different forms. Kinetic and gravitational potential energy are just two forms. Other forms include: heat or thermal energy, chemical potential energy, electrical energy, electromagnetic energy, sound energy, geothermal energy, wind energy, solar energy, and nuclear energy. The standard unit of energy in science is the joule. The unit was named in honor of James Prescott Joule, an English scientist who lived from 1818 to 1889. Joule was motivated by theological beliefs and wanted to show the unity of all the forces in nature. In the 1840s he determined the mechanical equivalent of heat by measuring the change in temperature produced by the friction of a paddle wheel attached to a falling weight.

Thermodynamics

Thermodynamics is the branch of physics that deals with the nature of heat and its conversions to other forms of energy.
The principle of the conservation of energy states that energy can neither be created nor destroyed. A body contains a total amount of energy as a result of the positions, motions, and chemical nature of its atoms and molecules. This is called internal or intrinsic energy. Bodies can change their internal energy by absorbing or giving off heat, by having work done on them, or by doing work. The first law of thermodynamics relates changes in internal energy to heat added to a system and the work done by a system. If a body undergoes a process of heat and work transfer, then the net heat supplied, plus the net work input, is equal to the change in internal energy.

The second law of thermodynamics is related to entropy. Entropy is a measure of disorder. The entropy of the universe increases. What this means is that an energy transfer is not a two-way street. It is impossible to convert all the heat from a given body into mechanical work.

Goals and Assessments

Goal	Location in Activity	Assessment Opportunity
Infer the loss of energy in the form of heat from the human body.	For You To Do Steps 1 – 7	For You To Do Step 7 Students are able to explain their observations.
Relate the laws of thermodynamics to the transfer of energy in a food chain.	BioTalk	Biology to Go, Questions 3 and 4.
Calculate the energy lost at a given level in a food web.	BioTalk	Biology to Go, Question 6.
Explain the significance of a pyramid of biomass, a pyramid of numbers, and a pyramid of energy.	BioTalk	Biology to Go, Question 5.

Activity Overview

In this activity students record and graph the temperatures in three containers of water. One container contains only water. A student places a hand in each of the other two containers. In one of these containers, the student holds the hand still and in the second container the student moves the hand. From the data students infer that heat is dissipated from the human body.

Preparation and Materials Needed

Preparation

You may have to freeze some ice a day before the class.

Materials/Equipment Needed (per group)

- 3 large containers (1000 mL)
- 3 thermometers
- ice
- stirring rod

(per student)
- graph paper

Learning Strategies for Students with Limited English Proficiency

1. Point out new vocabulary in context. Practice using the words as much as possible.

pyramid	trophic level	thermodynamics
dissipated	amplification	

2. In **Step 1** students are asked to read through the steps of the activity and predict what will happen. Ask the students to diagram what they will be doing in **Steps 2 to 6** as they read through the activity.

3. Presenting scientific data in graphical form is an important science skill. Look through the **Assessment Rubric for Graphs** provided in the Teacher's Edition at the end of this activity. Make sure that students understand what is meant by each rubric element. It is helpful for students to review each step separately, because some students may be learning the vocabulary of graphing, while others may be learning the content. Provide students with examples of graphs with missing information (such as titles or units on axes) or that have been incorrectly graphed (a typical student error is to place uneven intervals along an axis).

4. You may wish to dramatize the loss of energy at each level of a food chain. You will need to do this outdoors. Choose five students to represent five trophic levels. Provide each student with a large Styrofoam® cup. Use a pencil to poke holes in four of the cups. Ask the first student to stand beside a bucket full of water. The next three students should stand about 5 m apart from each other. The fifth student holds the cup without a hole and stands 5 m away from the fourth student. The first student begins by filling the Styrofoam cup with water, carrying it 5 m to the second student and then dumping the water into the second student's cup. The second student carries his cup 5 m to the third student and dumps the water into her cup. The third student carries her cup to the fourth student and dumps it into his cup. Finally, the fourth student carries his cup the 5 m to the fifth student and dumps it into the cup without a hole. Have the students compare the amount of water contained in the last cup to the amount in a full cup. You can also equate the bucket of water to the Sun, since only a small part of the available sunlight enters the food chain.

   If time permits, you can compare the amount of energy loss with fewer trophic levels.

A Vote for Ecology

Activity 3 Energy Flow in Ecosystems

GOALS

In this activity you will:

- Infer the loss of energy in the form of heat from the human body.
- Relate the laws of thermodynamics to the transfer of energy in a food chain.
- Calculate the energy lost at a given level in a food web.
- Explain the significance of a pyramid of biomass, a pyramid of numbers, and a pyramid of energy.

What Do You Think?

Heat stroke is caused by a failure of the heat-regulating mechanisms of the body. It can be caused by heavy exercise combined with hot and humid conditions.

- **Where does the heat in the body come from?**

Write your answer to this question in your *Active Biology* log. Be prepared to discuss your ideas with your small group and other members of your class.

For You To Do

As you work through this activity, consider whether there is any relationship between events like heat stroke and the heat that is stored and lost at each link in a food web.

1. Read through the steps of the activity.
   a) What are you investigating in this activity?
   b) Predict what you think will happen to the water temperatures in the containers.

What Do You Think?

Metabolic processes involve chemical reactions that release energy (exergonic reactions), and reactions that require energy to occur (endergonic reactions). Exergonic reactions provide for the energy needs of endergonic reactions. However, not all of the energy released by an exergonic reaction is free to be used by an endergonic one. There is always some "waste" energy in the form of heat. This is the source of the heat.

Student Conceptions

Students often think of energy as a thing. We are probably all responsible for this misconception, since we frequently will say, "close the door so you don't let the heat out (or the cold in)." With that comment we imply that heat is a thing, just as a cat and dog can be prevented from leaving by closing a door. This misconception may be further strengthened by talking about a flow of energy, leaving students to believe that energy is a type of fluid that is transferred in certain processes.

Some students may think of energy only in terms of activity. The fact that energy is defined as the ability to do work may contribute to this misconception. Students should be aware of the fact that although one form of energy, kinetic energy, is associated with movement, stored energy, or potential energy, is not associated with any type of activity.

For You To Do
Teaching Suggestions and Sample Answers

1. Although this activity begins by asking students to read through the steps, remind the students that they should always read through the steps of an activity before they begin. As suggested earlier, you may wish to ask students to diagram the steps of the activity.
   a) It is hoped that students will be able to understand that they are investigating heat that is dissipated from the human body.
   b) Student predictions will vary. Some may think that the water in the container with the moving hand may increase in temperature the most. Others may think that the movement of the hand may cause the temperature of the water in that container to decrease. Regardless of what they predict, be sure that they write down their prediction and that they provide an explanation.

2. You can now follow the steps to conduct the experiment. Put 600 mL of water in each of three containers. The temperature of the water should be 10°C. You may have to add ice. Remove the ice when the temperature gets to 10°C.
3. Have one student put one hand into the water in container A. Have that student put the other hand into the water in container B. In container A, move the fingers rapidly in the water. Do not move the hand in container B. Keep one hand moving and the other hand still for five minutes.
4. Another student will hold a thermometer in the water in container A. Read the temperature once each minute for 5 minutes.
   a) What is the purpose of stirring the water?
   b) Record the temperatures in the chart.

Clean up any spilled water immediately.

Minutes	Temperature Container A (moving hand)	Temperature Container B (still hand)	Temperature Container C (no hand)
1			
2			
3			
4			
5			

   a) Record the temperatures in a chart similar to the above.
5. A third student will hold a thermometer and read the temperatures in container B. Also, stir the water in this container using the stirring rod.

Wash your hands after completing the activity.

6. A fourth student will hold a thermometer in container C. Stir the water in this container. Read the temperature once each minute for five minutes.
   a) Why did you have a container that you did not put your hand in?

4. a) Students record their temperature readings.
   They should observe the temperature in container A rise the most.

Teaching Tip

The blank data table is available as
Blackline Master Ecology 3.1: Energy Flow Data.

5. Remind students not to use the thermometer to stir the water.
   a) Stirring the water keeps the temperature even throughout the container.
   b) Students should observe that the change in temperature in container B is less than that in container A.

6. a) Container C is the control. Point out to the students the need for controls in scientific experiments.

A Vote for Ecology

b) Record the temperature readings in the chart.

7. Make a line graph of the temperature readings for the three containers. You will have three lines on the same graph.

   a) In container B, you held your hand in cold water without moving it. What happened to the temperature? Does this data support your prediction?

   b) In container A, you exercised your hand. How did the temperature of the water change? Do your data support your prediction?

Pyramids of Mass and Energy

One of the most important abiotic factors that affects relationships in a community is energy. Organisms in an ecosystem are tied together by the flow of energy from one organism to another. The food chain that exists when a herbivore eats a plant and a carnivore eats a herbivore depends on the energy entering the community in the form of sunlight. Without the Sun, there would be no green plants, no herbivores, and no carnivores. (There are a few ecosystems that get their energy from another source.)

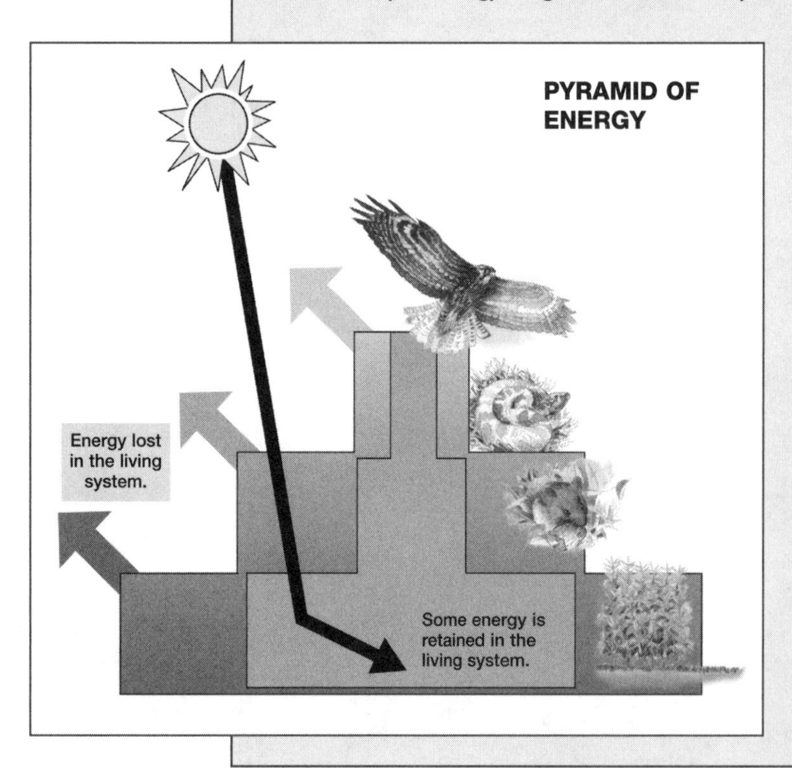

b) Students should observe the least change in temperature in container C.

7. Students graph their data.
   a) and b) The temperature increased in both containers, but increased more in container A. Have students examine their predictions at the beginning of the experiment in light of their experimental evidence.

Teaching Tip

You may wish to supplement this activity with the dramatization suggested in **Learning Strategies for Students with Limited English Proficiency.**

Assessment Opportunity

You may wish to use the **Assessment Rubric for Graphs** to assess the students' graphs. This rubric is available in the Teacher's Edition at the end of this activity.

BioTalk

Teaching Tip

The illustration on student page 554 is available as **Blackline Master Ecology 3.2: Pyramid of Energy.**

Teaching Tip

In 1977 geologists were exploring fractures in the ocean floor. Located deep in the ocean, they found thermal vents surrounded by a very unusual ecosystem. The vents spew out hydrogen sulfide. No one would have expected to find life around these vents. The water here is frigid, except near the vents, were it is incredibly hot–hundreds of degrees Celsius. No sunlight can penetrate that deep into the water, so there are no plants because they need sunlight to photosynthesize. The scientists were amazed to discover large tube worms, almost a meter in length. These worms had no mouths or digestive systems. They also found white crabs that were blind.

They discovered that the producers in this food chain were chemosynthetic bacteria. These bacteria used chemical energy in the hydrogen sulfide gas to make their food. The other organisms in this ecosystem ate the bacteria, harbored bacteria in their bodies, or ate the bacteria eaters.

The size of a community, therefore, is limited by the amount of energy entering it through its producers. The total amount of chemical energy stored by photosynthesis is the gross primary productivity of the community. Much of that energy is used by the producers to grow and to maintain themselves. The remaining energy, which is available to the consumers as food, is the net primary productivity of the community.

The transfer of energy from producer to primary consumer to secondary consumer, and so on in a food web must follow the laws of **thermodynamics**.

The first law of thermodynamics states that although energy can be transformed, it cannot be created or destroy. Some energy from the Sun is transformed into a form that can be used by living organisms. However, if energy is not destroyed, what happens to it? Why is it necessary to keep adding energy in the form of sunlight? That is where the second law of thermodynamics comes into play. It states that in any energy transformation some energy is lost from the system in an unusable form.

Usually this is in the form of heat. In this activity, you actually measured the temperature increases that resulted from the heat loss from the human body. You noted that with exercise, the heat loss was even greater than without movement.

Among living beings, the transfer of energy in food from "eaten" to "eater" is really quite inefficient, and of course a great deal of the food does not get eaten at all. From grass to sheep the loss is about 90 percent.

It takes about 10 kg of organic matter in the grass to support one kilogram of sheep.

Bio Words

thermodynamics: the study of energy transformations described by laws

NOTES

A Vote for Ecology

Bio Words

pyramid of living matter: a pyramid developed on the basis of the mass of dry living matter at each trophic level

pyramid of energy: a pyramid developed on the basis of the energy at each trophic level

trophic level: the number of energy transfers an organism is from the original solar energy entering an ecosystem; the feeding level of one or more populations in a food web

For the sake of simplification, assume that each consumer lives entirely on one kind of food. Then a person on a lake might live entirely on a given type of fish, for example. To support one kilogram of this person it takes about 10 kg of fish, 100 kg of minnows, 1000 kg of water fleas, and 10,000 kg of algae. This information in graph form is called a **pyramid of living matter**. Mass is a measure of the amount of matter in an object. Because much of the mass of living organisms is water, the producers first must be dried for a truer estimate of their mass when constructing a pyramid of matter. The pyramid shows that the amount of matter is greatest at the producer level.

It is possible to measure the amount of energy available at each level. The **pyramid of energy** that results from graphing these values also

THEORETICAL PYRAMID OF LIVING MATTER

human 100 kg
fish 1000 kg
minnows 10,000 kg
water fleas 100,000 kg
algae 1,000,000 kg

Humans have relied on fish as a source of food throughout history. Most of the fish protein was provided by species caught in the wild.

Teaching Tip

Students may wonder why the numbers in the text do not match the numbers in the diagram. Point out that the text talks about supporting 1 kg of human mass, where as the diagram represents the living matter required to support a person with a mass of 100 kg.

Note that the diagram is not really drawn to scale. You may wish to ask students to draw a diagram to scale to illustrate the tremendous size of the producer box, compared to the top consumer.

Assessment Opportunity

You may provide the students with the following questions to assess their understanding of the reading:

1. What is the source of energy that enters the food chain?
   (The Sun)
2. What limits the size of a community?
   (The energy provided by the producers)
3. State the first and second laws of thermodynamics.
   (The first law states that although energy can be transformed, it cannot be created or destroyed. The second law states that in any energy transformation, some energy is lost from the system in an unusable form.)
4. What form does the unusable energy lost from a system usually take?
   (Heat)
5. What is a trophic level?
   (Energy level)

shows that the energy available is greatest at the producer level and steadily decreases at the other levels. Each step in the pyramid is called a **trophic level** (energy level). Because energy is lost at each transfer, the steps in a pyramid of energy are limited. Usually, there are no more than about five trophic levels in a food chain.

It is also possible to construct a pyramid of numbers by counting the number of organisms in a food chain. Although the largest number of organisms is usually found at the base of the pyramid, this is not always the case. For example, in a meadow there will be many more grass plants than there will be grasshoppers. However, a single tree can sustain many caterpillars.

Reflecting on the Activity and the Challenge

In this activity, you observed the loss of heat from the human body. You then related that to the loss of energy at each step of a food chain. You learned that the further you go up a food chain, the less energy is available. As part of your challenge you are expected to explain how energy flows through an ecosystem. You should also consider how the flow of energy is affected in the environmental issue that you have chosen.

Biology to Go

1. What is the relationship, if any, between the heat energy stored and dissipated at each link in a food web and the heat energy responsible for a heat stroke?
2. In the activity, the student who kept his/her hand in the water may have begun to shiver. Why do you suppose this happened?
3. Explain how the transfer of energy in a food chain follows the laws of thermodynamics.
4. Why is there a limit to the number of trophic levels in an energy pyramid?
5. Why is a pyramid of numbers not always a good example of the flow of energy through a food chain?

Biology to Go

1. As energy is stored in a human during the dynamics of a food web some of it is dissipated as heat energy. This might be used to maintain normal body temperature or be given off into the environment. During exercise, cellular respiration speeds up to provide extra energy for the activity. A byproduct of this process is thermal energy and the excess may have to be eliminated to maintain normal body temperature. If the heat-regulating mechanisms of the body are unable to eliminate this excess heat produced by the high degree of physical activity because of a hot and humid environment the body temperature rises and could result in heat stroke.

2. Shivering is the same as any muscle activity; it results in an increase in heat energy. Hands in cold water cause the body to lose heat to the environment. Shivering is a protective action that tries to produce heat to replace the heat that is being lost. Violent shivering can increase the body's heat by as much as 18 times normal.

3. The first law of thermodynamics states that energy can be transformed, but cannot be created or destroyed. Energy from the Sun is transformed by plants into a form that can be used by the consumers in a food chain. The second law of thermodynamics states that in any energy transformation some energy is lost from the system. At each step of the food chain about 10% of the energy is lost.

4. Energy is lost at each step of a pyramid of energy. This puts a limit on the number of trophic or energy levels that can be sustained.

5. A pyramid of numbers is often not a good example of the flow of energy. In some cases a single tree can sustain a large number of first-order consumers.

6. An energy pyramid illustrates the energy lost at each level of a food web. In general, each level of the pyramid has only 10% of the energy at the level below it. If the producer level (the lowest level) has 10,000 kilocalories available for the rest of the food web, how much energy is available for the other three levels?

7. An energy pyramid illustrates a great loss of energy as you go up the pyramid. When humans eat meat, they act as a top-level consumer. A steer eats a small amount of corn which contains 10 kilocalories. If you were to eat the same amount of corn you would get the same amount of energy from it as the steer. How much energy would you get from that small amount of corn if you ate some hamburger from that steer? Before you calculate this answer think about how this energy pyramid differs from the energy pyramid in the previous questions.

Inquiring Further

Biological amplification

What does biological amplification mean? Use the example of dichloro diphenyl trichloroethane, or DDT, to illustrate how biological amplification and food chains and energy pyramids are related.

The peregrine falcon is a bird of prey at the top of the food chain. As a result of biological amplification, falcons ingested high levels of the pesticide DDT. Falcons contaminated with DDT did not lay eggs or produced eggs with shells that broke.

6. Producers = 10,000 kcals
   Primary consumers = 10,000 X 10% = 1000 kcals
   Secondary consumers = 1000 X 10% = 100 kcals
   Top-level consumers = 100 X 10% = 10 kcals

7. In this pyramid there are only three levels. There is no primary consumer.
   The corn, the producer, contains 10 kcals of energy.
   The steer, the secondary consumer, gets 10% of that energy, or 1 kcals.
   The human, the top consumer, get 10% of that energy, or 0.1 kcals.
   This illustrates how humans can obtain far more food energy by eating a given amount of grain than by eating the chicken or beef supported by the same amount of grain.

Inquiring Further

Biological amplification

Biological amplification is the tendency of pollutants to become concentrated in successive trophic levels. Often, this is to the detriment of the organisms in which these materials concentrate, since the pollutants are often toxic. DDT is a typical example of the effect biocides have on the relationships in a community. For example, at one time, marshes along the north shore of Long Island, New York were sprayed with DDT to control mosquitoes. Later, microscopic organisms in the marsh waters were found to have about 0.04 ppm (parts per million) of DDT in their cells. While this is a low level of poison, no one had expected to find any poison in the organisms. The minnows, clams, and snails that ate these organisms had levels of DDT more than 10 times higher, between 0.5 and 0.9 ppm. The eels, flukes, and billfish that ate the snails and small fish have levels of DDT ranging from 1.3 to 2.0 ppm. The ospreys, herons, and gulls that ate the eels and minnows had levels of DDT between 10 and 25 ppm. Thus, the concentration of DDT in the tissues of the organisms in this food chain increased almost 10 million times from the amount in the seawater and nearly 625 times from the produced level at the bottom of the pyramid to the consumer level at the top.

Blackline Master Ecology: Assessment Rubric: Graphing Skills

This assessment rubric breaks down skills in graphing data to give an idea where work is needed or where skills are lacking. For each element, place a check mark in the appropriate box to indicate a score of 0, 1 or 2. The maximum score is 20 points if the rubric is used to score graphs of more than one set of data, or 18 points if only one set of data (when element 10 is not applicable).

Assessment Element	0 = Incorrect or not done	Score 1 = Work has errors or critical details are overlooked	2 = Thorough job without errors
1. Graph has a title that is relevant and informative.			
2. The variables plotted on the vertical and horizontal axes are clearly labeled.			
3. Units of measurement are provided in parentheses beside the axis labels.			
4. Independent variable is plotted along the horizontal axis. Dependent variable is plotted along the vertical axis.			
5. Appropriate ranges are chosen for axes.			
6. Appropriate intervals are chosen for axes and numbers are marked along axes.			
7. Data are plotted correctly.			
8. A line is drawn through the data points (either best-fit or connecting line, depending on instructions).			
9. Graph is neat and orderly, can be read with ease.			
10. If two or more sets of data are plotted on the same graph, they are shown in different colors. A legend is provided to indicate which data are which.			

Blackline Master Ecology 3.1: Energy Flow Data

Minutes	Temperature Container A (moving hand)	Temperature Container B (still hand)	Temperature Container C (no hand)
1			
2			
3			
4			
5			

Blackline Master Ecology 3.2: Pyramid of Energy

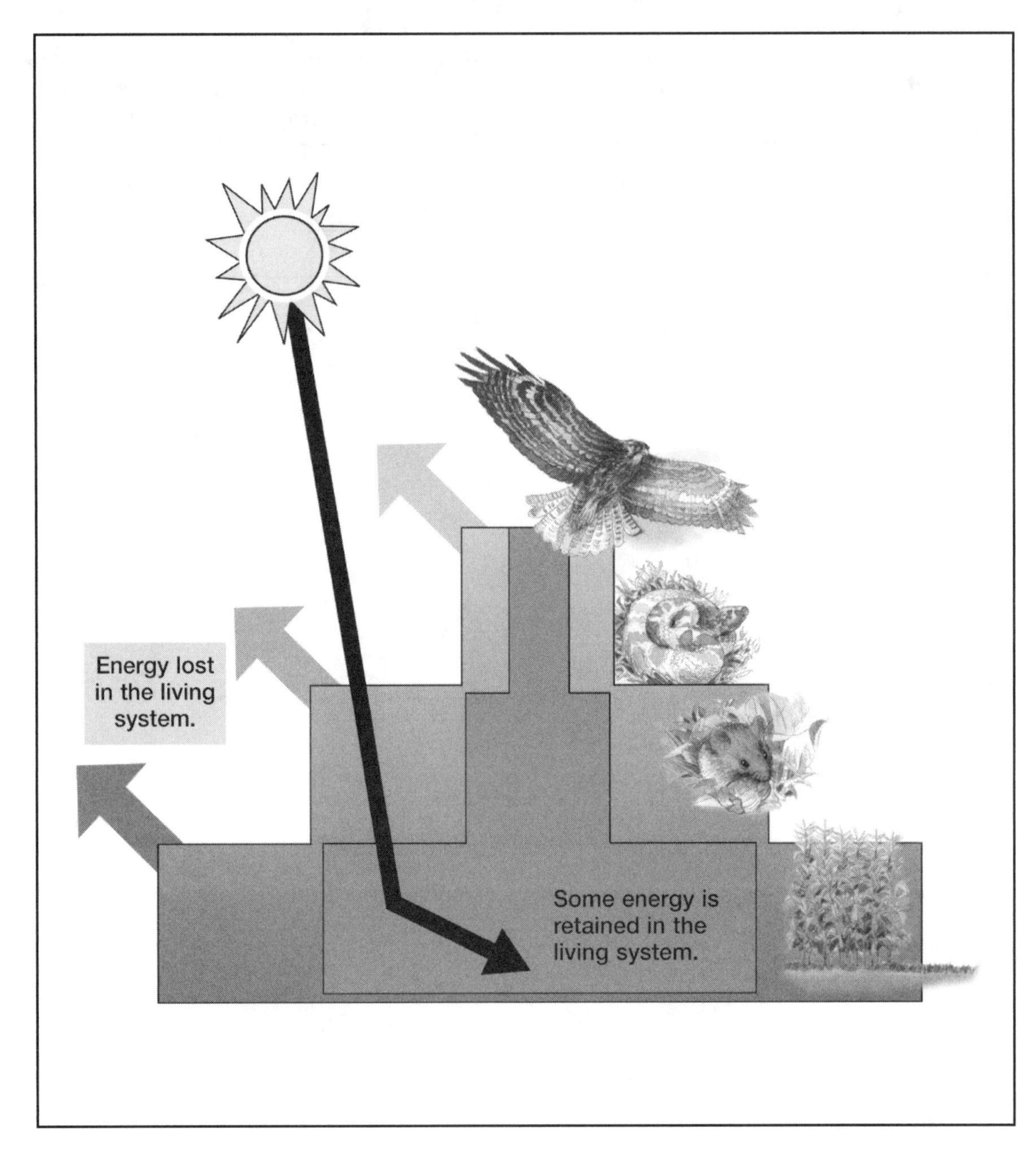

NOTES

ACTIVITY 4– FACTORS AFFECTING POPULATION SIZE

Background Information

Limiting Factors

In nature, a population varies between some upper and lower limit, depending upon the birthrate, death rate, immigration, and emigration rates. (These are explained in the **BioTalk**.) Each population is like a swing moving back and forth between two points, or between a high and low number. The lower limit, of course, is zero. At that point the population no longer exists. What is the upper limit on population size?

It depends on the environment and everything that surrounds and affects an organism. The environment may slow the individual's growth, may kill the individual, or may stimulate the individuals growth and reproduction. In any case, the environment affects individuals and, thus, the ultimate size of a population.

The environment is made up of two parts: the biotic and the abiotic. Both biotic and abiotic factors affect the size of a population. Any biotic or abiotic factor that can slow the growth of a population is called a limiting factor. The climate is a group of limiting factors that affect all plant and animal life. Water is another important limiting factor. All living organisms need water. Another abiotic limiting factor is space. The amount of space needed by a single organism is tied in part to a biotic factor–the availability of food energy.

Goals and Assessments

Goal	Location in Activity	Assessment Opportunity
Investigate the factors that affect the size of a population.	**For You To Do** Part A **BioTalk**	Students answer question in Part A Step 2 correctly. **Biology to Go** Question 1.
Interpret a graph and make calculations to examine factors affecting fluctuations in populations.	**For You To Do** Part B Human Population Steps 1 to 5	Students answer questions in Part B Steps 2 to 4 correctly.
Distinguish between an open and closed population.	**BioTalk**	**Biology to Go** Question 3.

Activity Overview

In this activity students compare the growth of a reindeer population on an island near Alaska and the human population. They examine the roles played by carrying capacity, doubling time, and the four rates that determine a population's size–birthrate, death rate, immigration, and emigration.

Preparation and Materials Needed

Preparation

Materials/Equipment Needed (per group)

- calculator

(per student)

- graph paper

Learning Strategies for Students with Limited English Proficiency

1. Point out new vocabulary in context. Practice using the words as much as possible.

population	fluctuation	reindeer	death rate (mortality rate)
cyclically	emigration	growth rate	birthrate (natality)
immigration	carrying capacity		

2. To help students understand the difference between exponential growth (2, 4, 8, 16, 32...) and arithmetic growth (1, 2, 3, 4, 5...) you may wish to remind them of the student that convinced his parents that they can pay his allowance in pennies. The first week he would take one penny, and each week after they would double the number of pennies. That is, the second week he would get two pennies, the third week he would get four pennies, and so on. Ask them how they would prefer to get paid–using the penny method or at $50 per week for one year. Then ask the students to calculate how much they would get at the end of a year using both methods. Explain that this is the difference between exponential growth and arithmetic growth. You may choose to refer to exponential growth simply as doubling.

3. Numbers such as those in the thousands of millions (billions) are difficult for students and most adults to comprehend. You may wish to share some of the following examples with your students:

   In his book called *How Much Is a Million?* David Schwartz shares this observation: "How big is a billion? If a billion kids made a human tower, they would stand up past the moon. If you sat down to count from one to one billion, you would be counting for 95 years. If you found a goldfish bowl large enough to hold a billion goldfish, it would be as big as a stadium."

   An advertising agency shared these calculations:
   - A billion seconds ago, it was 1964.
   - A billion minutes ago, Jesus was alive.
   - A billion hours ago, our ancestors were living in the Stone Age.
   - A billion dollars ago was only 8 hours and 20 minutes, at the rate Washington spends it.

   You may wish students to add to this list with their own calculations.

Activity 4 Factors Affecting Population Size

GOALS

In this activity you will:

- Investigate the factors that affect the size of a population.
- Interpret a graph and make calculations to examine factors affecting fluctuations in populations.
- Calculate the doubling time of the human population.
- Distinguish between an open and closed population.

What Do You Think?

The population of your community may be going up, going down, or remaining the same. The change depends on whether individuals are being added to or taken away from your community.

- **What can take place in your community, or any other community of living things, that can influence the size of the population?**

Write your answer to this question in your *Active Biology* log. Be prepared to discuss your ideas with your small group and other members of your class.

For You To Do

This activity provides an opportunity for you to examine the factors that affect the changes (fluctuations) that occur in a population in an ecosystem.

What Do You Think?

Four rates–natality, mortality, immigration, and emigration –determine a population's size.

Student Conceptions

Students who have lived in one area all their lives may not associate immigration and emigration with changes in the size of a population. Also, although students may understand what happened to the reindeer population on St. Paul Island, they may not appreciate its similarity to the human population.

A Vote for Ecology

Part A: Reindeer Population

1. In 1911, 25 reindeer, 4 males and 21 females, were introduced onto St. Paul Island near Alaska. On St. Paul Island there were no predators of the reindeer, and no hunting of the reindeer was allowed. Study the graph shown below and answer the questions in your *Active Biology* log.

   a) In 1911 the population was 25 reindeer. What was the size of the population in 1920? What was the difference in the number of reindeer between 1911 and 1920? What was the average annual increase in the number of reindeer between 1911 and 1920?

   b) What was the difference in population size between the years 1920 and 1930? What was the average annual increase in the number of reindeer in the years between 1920 and 1930?

   c) What was the average annual increase in the number of reindeer in the years between 1930 and 1938?

   d) During which of the three periods 1911—1920, 1920—1930, or 1930—1938, was the increase in the population of reindeer greatest?

   e) What was the greatest number of reindeer found on St. Paul Island between 1910 and 1950? In what year did this occur?

   f) In 1950, only eight reindeer were still alive. What is the average annual decrease in the number of reindeer in the years between 1938 and 1950?

2. In your group, discuss the questions on the next page. Then answer them in your *Active Biology* log.

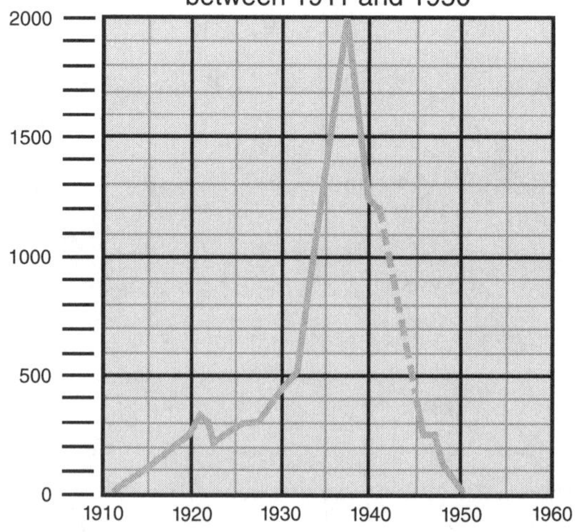

For You To Do

Teaching Suggestions and Sample Answers

Part A: Reindeer Population

1. For this step, suggest that students read the numbers from the graph to the nearest 100.
    a) In 1920 the size of the population was about 300. The difference between 1911 and 1920 was +250. The average annual increase was about 28.
    b) In 1930 the size of the population was about 500. The difference between 1920 and 1930 was +200. The average annual increase was about 20.
    c) In 1938 the size of the population was 2000. The difference between 1930 and 1938 was +1500. The average annual increase was about 187.
    d) The increase was the greatest from 1930 to 1938.
    e) The greatest number was 2000 in 1938.
    f) The difference between 1938 and 1950 was +1992. The average annual decrease was 166.

Teaching Tip

The graph on student page 560 is available as **Blackline Master Ecology 4.1: Changes in Reindeer Population.**

a) St. Paul Island is more than 323 km (200 miles) from the mainland. Could leaving or arriving at the island have played a major role in determining the size of the reindeer population? Explain your answer.

b) What might account for the tremendous increase in the population of reindeer between 1930 and 1938, as compared with the rate of growth during the first years the reindeer were on the island?

c) St. Paul Island is about 106 km^2 (41 square miles). What effect might 2000 reindeer have on the island and its vegetation?

d) Consider all the factors an organism requires to live. What might have happened on the island to cause the change in population size between 1938 and 1950?

e) Beginning in 1911, in which time spans did the reindeer population double? How many years did it take each of those doublings to occur? What happened to the doubling time between 1911 and 1938?

f) If some of the eight reindeer that were still alive in 1950 were males and some females, what do you predict would happen to the population in the next few years? Why?

g) What evidence is there that the carrying capacity (number of individuals in a population that the resources of a habitat can support) for reindeer on this island was exceeded?

h) What does this study tell you about unchecked population growth? What difference might hunters or predators have made?

Part B: Human Population

1. On a piece of graph paper, plot the growth of the human population using the following data.

Human Population Growth between A.D. 1 and 2000

Date A.D.	Human Population (millions)	Date A.D.	Human Population (millions)
1	250	1930	2070
1000	280	1940	2300
1200	384	1950	2500
1500	427	1960	3000
1650	470	1970	3700
1750	694	1980	4450
1850	1100	1990	5300
1900	1600	2000	6080
1920	1800	2010	?

2. a) Probably not. St. Paul Island is more than 323 km from the mainland. Reindeer are strong swimmers, but the distance is too great for emigration or immigration to have a major effect.
   b) Population growth is exponential. Exponential growth is characterized by doubling. A few doublings lead to enormous numbers. With a large population base to start with in 1930, the growth between 1930 and 1938 would be great.
   c) Overgrazing, death of plants, destruction of habitats, accumulation of wastes.
   d) Overgrazing resulted in the death of plants and insufficient food. Weakened by a lack of food, the reindeer were prey to disease, and the reproductive rate could have declined drastically.
   e) The population doubled in 1912 (1 year), 1915 (3 years), 1920 (5 years), 1930 (10 years), 1934 (4 years), 1937 (3 years). Doubling time between 1911 and 1938 became longer until 1930 and then shorter.
   f) The population might die out because it is too small to recover, or it may slow the increase. Rationales will vary.
   g) The population crested after 1938.
   h) Natural controls take effect and can have drastic results, such as the total population dying out. Predators and hunters might have controlled the population, preventing exponential growth and the destruction of the environment, thus maintaining the carrying capacity of the environment.

Teaching Tip

Graph paper is available as **Blackline Master Ecology 4.2: Graph of Human Population Growth.**

Assessment Opportunity

You may wish to use **Assessment Rubric for Graphs** available at the end of **Activity 3** in this Teacher's Edition to assess the students' graphs.

A Vote for Ecology

2. Use your graph to determine the doubling times for the human population between A.D. 1 and 2000.

   a) How much time elapsed before the human population of A.D. 1 doubled the first time?

   b) Is the amount of time needed for the human population to double increasing or decreasing?

   c) What does that indicate about how fast the human population is growing?

3. Extend your graph to the year 2010.

   a) What do you estimate the human population will be in that year?

4. Using the equations below, estimate the doubling time for the current population based on the rate of growth from 1990 to 2000.

   a) In what year will the present population double?

   c) In what ways is the Earth as a whole similar to an island such as St. Paul? Does the Earth have a carrying capacity? Explain your answer.

$$\text{Annual rate of growth (in percent)} = \frac{(\text{population in 2000} - \text{population in 1990}) \times 100}{\text{population in 1990} \times \text{number of years}}$$

$$\text{Doubling time} = \frac{70}{\text{annual rate of growth}}$$

5. In your group, discuss the following questions. Then answer them in your *Active Biology* log.

   a) What similarities do you see between the graph of the reindeer population and your graph of the human population?

   b) What are the three or four most important factors required to sustain a population?

   d) What might happen to the population of humans if the present growth rate continues?

   e) What methods could be used to reduce the growth rate?

   f) Suggest several problems in the United States that are related to the human population.

   g) What are the most important three or four factors to think about with regard to the world population?

Part B: Human Population

Human Population Growth between A.D. 1 and 2000			
Date A.D.	Human Population (millions)	Date A.D.	Human Population (millions)
1	250	1930	2070
1000	280	1940	2300
1200	384	1950	2500
1500	427	1960	3000
1650	470	1970	3700
1750	694	1980	4450
1850	1100	1990	5300
1900	1600	2000	6080
1920	1800	2010	?

Plot Date on *y*-axis and Human population (millions) on *x*-axis.
Title of graph is **Human Population Growth.**

2. a) 1660 years
   b) Decreasing
   c) The human population is growing exponentially (very rapidly).

3. a) About 7000 million

4. a)

$$\text{Annual rate of growth} = \frac{(\text{pop. in 2000 - pop. in 1990}) \times 100)}{\text{pop. in 1990} \times \text{number of years}}$$

$$= \frac{(6080 - 5300) \times 100}{\text{p5300} \times 10}$$

$$= 1.5\ \%\ \text{(approx.)}$$

$$\text{Doubling Time} = \frac{70}{1.5}$$

$$= 4.7$$

$$2000 + 47 = 2047$$

5. a) Both show exponential growth curves.
   b) Students may suggest some of the following factors:
   food, crop land, grazing land, forests, water, air.
   c) Earth is a finite environment, like an island in space.
   Yes, Earth has a carrying capacity because it has finite resources.
   d) The human population may decrease because of starvation or disease.
   e) Student answers will vary. Some suggested methods include: birth control, abortion, infanticide, restriction on the number of children parents may have.
   f) Answers may include: immigration, increasing water and air pollution, decreasing groundwater, increased soil erosion, decreased soil nutrients, loss of wildlife and wildlife habitat.
   g) Factors that the students may consider include:
   water, living space, food supply, quality of life, justice.

BioTalk

CHANGING POPULATION SIZES

Four Rates Determine Population Size

The size of a population changes through time. Suppose a biologist counted 700 ponderosa pines on a hill in Colorado in 1990. In 2000, when the biologist counted the trees again, there were only 500. In other words, there were 200 fewer trees in 2000 than in 1990.

There are many reasons that a population of trees may decrease. These include forest fires and logging. What else may contribute to the decrease of population?

This is a decrease in the population of ponderosa pines. This change in population may be expressed as a rate. To find the rate you divide the amount of change by the amount of time for the change to take place. The rate is an average. In this example, the rate of change in the number of trees divided by the change in time may be expressed as: -200 trees $\div$ 10 years $= -20$ trees per year. To the biologist, this means each year there were 20 fewer trees in the population. Keep in mind, however, that this rate is an average. It is unlikely the trees disappeared on a regular schedule. All of the trees may have been lost in one year due to a fire. Alternatively, selective cutting during several years may have caused the decrease.

BioTalk

Teaching Tip

Provide students with other examples of growth rate to illustrate both positive and negative growth rates. For example: the deer population in an area was 75 in 2000. In 2005, the population was 85.
What was the average rate of growth? *(85 – 75 = 10; 10/5 = 2)*

Ask the students to speculate how biologists might go about counting the trees on a hillside. *(They probably estimated by counting the number of trees in several sample areas and then extrapolating the information.)*

A Vote for Ecology

Bio Words

death rate (mortality rate): the rate at which death decreases the size of a population

birthrate (natality): the rate at which reproduction increases the size of a population

immigration: the number of individuals of a species that move into an existing population

emigration: the number of individuals of a species that move out of an existing population

growth rate: the rate at which the size of a population increases as a result of death rate, birthrate, immigration, and emigration

What does the decrease of 200 pine trees in 10 years represent? Because pine trees cannot wander away, they must have died or have been cut down. In this situation, then, the decrease represents the **death rate**, or **mortality rate**, of the pine population. The number of deaths in the pine population per unit of time is the mortality rate. Mortality is not the only change that can affect a population, however. While some of the pines may have died, some young pine trees may have started to grow from seed. Death decreases a population; reproduction increases it. The rate at which reproduction increases the population is called the **birthrate**, or **natality.**

Organisms that can move have two other ways to bring about a change in population size. Suppose you were studying the pigeon population in your city or town. You might discover that a certain number of pigeons flew into the city in one year. This is called **immigration**. It occurs when one or more organisms move into an area where others of their type can be found. Immigration increases the population. While studying the pigeons, you might notice that a certain number flew out of the city. This is called **emigration**. It occurs when organisms leave the area. Emigration decreases the population. In any population that can move, then, natality and immigration increase the population. Mortality and emigration decrease the population. Thus, the size of any population is the result of the relationships among these rates.

The number of individuals of a species that move into and out of an area will affect the size of a population.

Natality, mortality, immigration, and emigration rates apply to every population, including the human population. The sum of these rates makes up the **growth rate** of a population. The growth rate of a population is the number of organisms added to (or subtracted from) a population in a year due to natural increase and net migration. Often, this rate is expressed as a percentage of the population at the beginning of the time period.

NOTES

Population Density May Fluctuate

Any population has a built-in, characteristic growth rate. This is the rate at which the group would grow if food and space were unlimited and individuals bred freely. Environmental factors do affect a population's growth rate, however. The interaction of the population's natural growth rate and the environment determines the density of the surviving population. The maximum number of individuals that a given environment can support is called the **carrying capacity**.

Although there is variation among species, female ducks lay about 10 eggs per nesting attempt. The overall strategy for these birds is to get as many eggs out there as they can in the hopes that at least some will make it.

If you measure the density of a population at intervals during a given period of time, you seldom find any two consecutive measurements the same. Density increases or decreases continually. Most natural populations are **open populations**. These are populations in which individuals are free to emigrate or immigrate and in which the birth and death rates fluctuate. Variables in the environment, such as climate, available food, or the activities of natural enemies, are the causes of the fluctuations. In a closed population, birthrate and death rate are the only factors that affect the size of the population. The island of reindeer you studied in **Part A** is an example of a closed population.

Bio Words

carrying capacity: the maximum population that can be sustained by a given supply of resources

open population: a natural population in which all four factors that affect population size (death rate, birthrate, immigration, and emigration) are functioning

NOTES

Sometimes population fluctuations are fairly regular, and the peaks are at approximately equal time intervals. For example, populations of lemmings often peak every three or four years. Many of the animals that live in the northern parts of Europe, Asia, and North America show similar population cycles. Although the data show very regular cycles when they are plotted on a graph, the reasons for the seemingly regular cycles are not well understood. A combination of purely chance events also can produce apparently regular cycles.

Lemmings are known for repeated population explosions. During the peak, the population may increase a thousand times. Food becomes scarce and lemmings must migrate to new areas.

Although populations may change cyclically, many population changes are permanent. If a population becomes extinct, for example, the change is permanent. Any permanent change in a population is a change in the community to which the population belongs. Permanent changes in one population also may affect other populations of organisms in the same community.

Reflecting on the Activity and the Challenge

In this activity you discovered that birthrate, death rate, immigration, and emigration affect the growth rate of a natural, open population. Review the issue you have identified for research. Consider if any of the factors involving population size are relevant to the issue. You will be asked to explain to the public the importance of these factors in providing the science behind your stand.

Teaching Tip

Emphasize the concept of unit area in defining a growth in population. It may be a small area such as the water droplet on moss, or a larger area such as a pond, or an area as large as the Earth itself.

You may want to introduce density calculations at this time. The rate at which the density of a population changes is equal to the change in density divided by the change in time.

Ask students why, when considering carrying capacity, we must think of the whole Earth for humans but not for pigeons or pine trees. *(Pines and pigeons are widely spread over the Earth, but they are not as ubiquitous as people. Moreover, human activities do not affect only the parts of the biosphere humans inhabit but all parts, for example the stratosphere and the oceans.)*

Assessment Opportunity

You may provide the students with the following questions to assess their understanding of the reading.

1. How do you calculate the rate of change in a population?

   (You divide the amount of change by the amount of time for the change to take place.)

2. What are the four limiting factors that affect a population?

   (birthrate, death rate, immigration, and emigration)

3. What does carrying capacity mean?

   (The maximum number of individuals that a given environment can support)

4. What is the difference between an open and a closed population?

   (In an open population all four factors that affect population size are functioning. In a closed population, only birthrate and death rate affect population size.)

Chapter 9

Biology to Go

1. How do each of the four limiting factors affect population growth?
2. Explain how limiting factors could play a role in the extinction of a population.
3. Distinguish between an open and closed ecosystem. Use examples to illustrate your answer.
4. Scientists studying an area of the tundra reported that they found 5 lemmings per hectare. They returned the following year and discovered that the density of the lemmings in the same area were 20 per hectare. What is the rate of growth of lemmings in the area, expressed as a percentage?
5. According to the U.S. Census Bureau, the population of the United States is influenced by the following:

   1 birth every 8 s

   1 death every 13 s

   1 immigrant every 22 s.

   Use these figures to determine the time, in seconds, it takes for the net gain of one person. (Hint: Start by calculating the number of births, deaths, and immigrants every minute. Round off to whole numbers.)

Inquiring Further

1. Population growth in different parts of the world

Research a place in the world where population growth is a problem today. How is it a problem? Research a place in the world where population growth is not a problem today. Why is it not a problem?

2. The truth behind lemming suicide

During the filming of the 1958 Disney nature documentary *White Wilderness*, the film crew induced lemmings into jumping off a cliff and into the "sea" in order to document their supposedly suicidal behavior. Research and report on the truth of this statement and the truth about lemming "suicide."

Biology to Go

1. Birth and immigration increase a population and death and emigration decrease a population.

2. If the death rate were greater than the birthrate over an extended period of time, a population would become extinct.

3. An open ecosystem is one in which all four factors: birthrate, death rate, immigration, and emigration are operating. Students can provide an example of any natural ecosystem like one operating in a forest. A closed ecosystem is one in which only birthrate and death rate are operating. The island of reindeer was a closed ecosystem.

4. Annual rate of growth = (20 – 5) × 100)
   = 5 × 1
   = 300 %

5. Births: 60 divided by 8 = 7.5 or about 8 births every minute
   Deaths: 60 divided by 13 = 4.6 or about 5 deaths every minute
   Immigrants: 60 divided by 22 = 2.7 or 3 migrants every minute
   8 births + 3 immigrants = 11 persons every minute
   11– 5 deaths = a net gain of 6 persons every minute
   60 seconds divided by 6 = 1 person every 10 seconds.

Inquiring Further

1. **Population growth in different parts of the world**
   In South America, Central America, Africa, and India population growth is a problem. The birthrates here are high, resulting in insufficient food and poor economic conditions. In Northern Europe the birthrates are low, and the overall quality of education is high.

2. **The truth behind lemming suicide**
   The statement about the Disney film crew is true. However, there is no truth behind lemming suicide.

Blackline Master Ecology 4.1: Changes in Reindeer Population

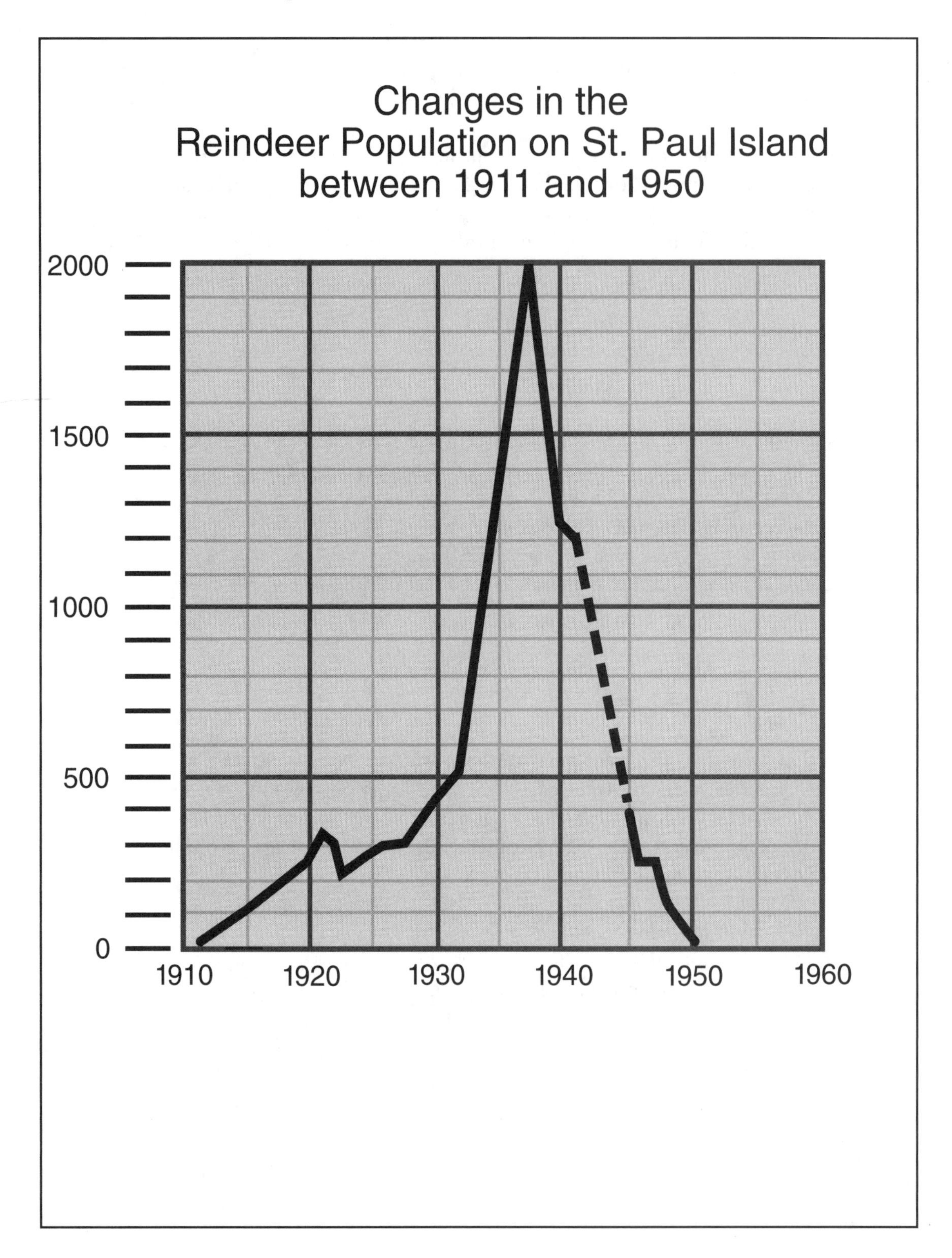

Blackline Master Ecology 4.2: Graph of Human Population Growth

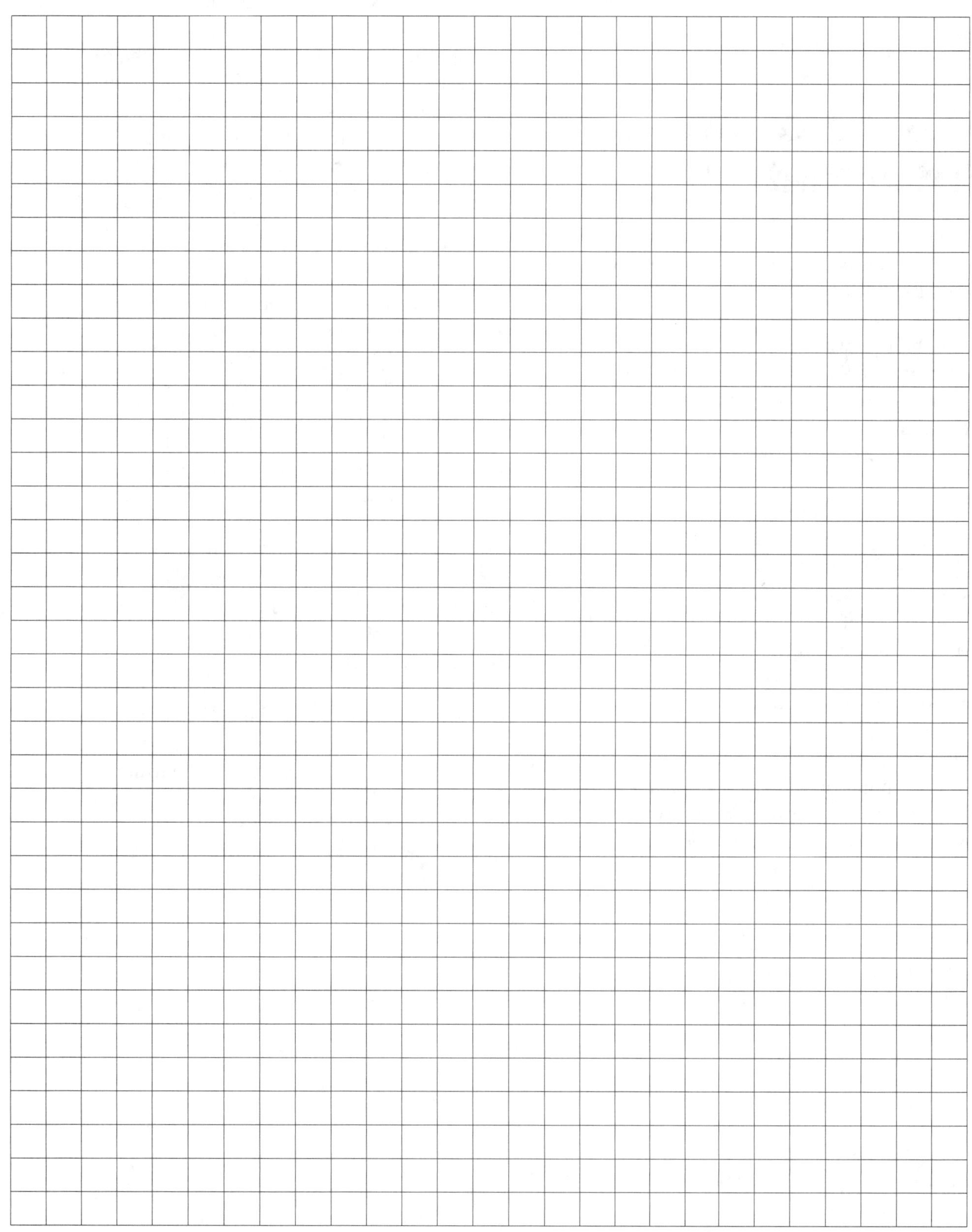

ACTIVITY 5– COMPETITION AMONG ORGANISMS

Background Information

Competition

Competition occurs between two or more organisms. It occurs when the use of a resource by one organism results in a reduced availability of the resource for other organisms. This limits the potential growth and reproduction of the other organisms. A resource is any factor or material that is required for survival, growth, and reproduction. Although competition occurs at the individual level, the repercussions are felt by the entire population. There are two types of competition. When competition occurs among organisms that belong to the same species it is referred to as **intraspecific** competition. When it occurs between individuals of different species it is **interspecific** competition.

Intraspecies competition varies as the density of the population of organisms changes. For example, as the size or density of a opulation increases, so does the intensity of the competition. As the size or density of a population decreases the strength of the intraspecies competition decreases as well. This acts to stabilize a population.

Interspecific competition, on the other hand, can destabilize a population. This can occur if one species is less capable of competing for the limiting resource. The strength of the competition in this situation is determined by the stronger competitor and may eventually lead to the extinction of the weaker competitor. Two species with very similar resource requirements cannot exist in the same place at the same time if one competitor is stronger than another.

There are two ways in which species can compete. One is through exploitation. (Also referred to as scramble competition.) In this type of competition the two species differ in their abilities to use or extract the resource. One is more efficient than the other. For example, the feeding rates in animals might differ, or the ability of one plant to extract water from the soil may differ from that of other plants. The other way is through interference competition or contest competition. In this type of competition one organism directly affects the ability of the other organism to obtain a resource by physical contact. This is done by guarding or defending the resource by displays or fighting. It can also be accomplished by chemical means like allelopathy.

Goals and Assessments

Goal	Location in Activity	Assessment Opportunity
Observe the effects of competition among plants for space and nutrients.	**For You To Do** Parts A and B	Student's responses to questions in Part A Step 8 and Part B Step 4 are appropiate.
Describe the possible effects of introducing a nonnative species into an ecosystem.	**BioTalk**	**Biology to Go** Question 3.
Explain why competition in nature is important.	**BioTalk**	**Biology to Go** Questions 2 and 4.

Activity Overview

Students observe the difference in the growth of plants when the plants must compete for space and nutrients. They also observe what happens when different species compete. They then design and carry out an experiment to study competition under different conditions.

Preparation and Materials Needed

Preparation

You and the students will need to set up **Part A** of this activity at least three weeks ahead of the time that you are ready to discuss competition. Students will be making and recording observations approximately every two days during the three weeks. You do not need to discuss the importance of competition prior to planting the seeds. Students will still be able to make meaningful observations. Leave the **What Do You Think?** questions until you are ready to teach the lesson.

You may also wish to set up **Part B** in advance. However, it would also be appropriate for students to do this part of the activity as an extension. Discussion of adaptations continues in the following chapter on evolution and experimental results would be equally meaningful at that time.

Materials/Equipment Needed (per group)

- 6 milk container bases (or similar containers)
- potting soil
- cress seeds (about 80)
- lettuce seeds (about 15)
- variety of other seeds (students' choice)
- small watering can (or suitable substitute)
- ruler

Learning Strategies for Students with Limited English Proficiency

1. Point out new vocabulary in context. Practice using the words as much as possible.

   competition nonnative invasive

2. Although only a couple of new vocabulary words are introduced in this activity, some of the reading in the **BioTalk** may challenge students with limited English proficiency. The following sentences in the Student Edition on page 573 can be illustrated with pictures:

   "The spread of nonnative organisms destroys healthy, diverse ecosystems. It replaces them with biologically impoverished, homogeneous landscapes."

   Have students search for photographs to illustrate each type of ecosystem. Ask them to explain, using the photos, what makes one ecosystem healthy and the other impoverished and homogeneous.

A Vote for Ecology

Activity 5 Competition Among Organisms

GOALS

In this activity you will:

- Observe the effects of competition among plants for space and nutrients.
- Describe the possible effects of introducing a nonnative species into an ecosystem.
- Explain why competition in nature is important.

What Do You Think?

Nature documentaries often feature the competition among animals for food, water, and space. These scenes are exciting to watch. However, plants seem to take a backstage to this type of activity.

- **Do plants need to compete among themselves in any given environment?**
- **If plants do compete, how do they do it?**

Write your answer to these questions in your *Active Biology* log. Be prepared to discuss your ideas with your small group and other members of your class.

For You To Do

This activity gives you an opportunity to observe the effect on plant growth when plants must compete for nutrients and space.

What Do You Think?

- Plants compete among themselves for moisture, inorganic nutrients, space, light, and air. Resources in any ecosystem are limited.

- Plants compete through the efficiency through which they uptake and use water and by their root distribution, both vertical and horizontal.

Student Conceptions

Students will be familiar with competition among organisms in the form of predation in animals. The **What Do You Think?** questions ask the students to consider competition among plants. Also, stress that competition can occur among microorganisms as well.

Part A: Competition among Plants

1. Fill five milk container bases with soil. Label the containers A through E.
2. Thoroughly moisten the soil in each container. Use the same amount of water to moisten the soil in each container.
   a) Why is it necessary to use the same amount of water in each container?
5. Place your containers in a low-light, room-temperature location (20°C is optimal). Keep the soil moist, but not soggy, by watering or misting every day or two.
6. When your plants sprout and begin to shed their hulls they are ready for light. Move them to a well-lighted location. If you go with sunlight be prepared to water more frequently. Room light will work as well and will not dry out the soil as quickly.

Do not eat any of the seeds in this activity. Wash your hands after handling the seeds and the soil.

3. Place seeds into the containers as follows:
   A. 5 crinkly cress (pepper grass) seeds
   B. 10 crinkly cress seeds
   C. 20 crinkly cress seeds
   D. 30 crinkly cress seeds
   E. 15 crinkly cress and 15 lettuce seeds.

   Spread seeds on the thoroughly moistened soil. (You are not expected to place them one at a time. Just spread them out as much as you can and as evenly as you can.)
4. Cover the seeds with a thin layer of soil.
   a) Count and record the number of seeds that germinated.
7. Keep the soil moist but not soggy by watering regularly. Water from the side if possible to prevent injuring the tiny plants. Again, make sure that each container is given the same amount of water.
8. Measure the heights of the plants and note the number and size of the leaves every two days for about three weeks.
   a) Record your observations.
   b) In which container were the cress plants the tallest?

For You To Do
Teaching Suggestions and Sample Answers

Part A: Competition among Plants

1. You may wish to substitute the milk containers with plastic flower pots of a similar size. They are relatively inexpensive and can be reused.

Teaching Tips

This is a good opportunity to remind students of the importance of controlling variables in any experiment.

You may wish to substitute the seeds suggested for this activity with rapid-growing plant seeds.

2. a) All variables except the one being studied must be controlled.

4. A sieve is handy to use to spread a thin and even layer of soil over the seeds.

6. a) You can expect that some seeds will not germinate.

Teaching Tip

You may wish to have the students make quantitative observations by measuring the height of the plants. If you do, this is also an excellent chance to introduce or remind students of the uncertainty in measurements.

8. a) Student observations will vary.
   b) The cress plants will be tallest in the container with only five seeds.

c) In which container did the cress plants have the greatest number of leaves?

d) What happened to the appearance of the cress plants as the number of seeds in a container was increased?

e) Account for your observations when cress and lettuce seeds were planted together.

In the next part of this activity you will determine which plant species has an advantage under certain conditions. Each research group within the class can study a different set of variables.

Part B: Competition under Different Conditions

1. In this part of the activity you will once again use the bases of milk containers with moist potting soil to plant your seeds.
2. Plant seeds from various species according to the instructions on the packets. As a class, decide how you will control the mix of seeds in each container.
3. Water each of the milk cartons with the same amount of water every second day.

   a) Record the amount of water used.
4. Once seeds start to germinate, store each of the milk cartons in a different environment. You might want to use temperature, amount of light, or amount of water as variables. Measure the growth of each of the plants daily.

   a) Does one type of plant begin to dominate the community? Is it the same type of plant in all containers?

   b) Present your data and draw a conclusion.

   c) Speculate about why one plant might be better adapted for a specific environment.

Do not eat any of the seeds in this activity. Wash your hands after handling the seeds and soil.

c) The cress plants will have the greatest number of leaves in the container with only five seeds.
d) As the number of cress plants in a container increased, each individual plant appeared to be smaller.
e) Cress seeds planted together with lettuce seeds is an example of interspecies competition.

Part B: Competition under Different Conditions

Teaching Tip

Factors that students might consider include such things as temperature, nutrients, light, soil characteristics (other than nutrients), and moisture.

3. Students may choose to vary the amount of water, in which case they do not need to water with the same amount of water. However, they must still record the amount of water used.

4. a) and b) Data will depend on the type of seeds used and the conditions chosen to investigate.
   c) Students should come to realize that plants have adapted to the environments in which they live. Therefore, a plant that naturally grows in the shade will thrive in a low-light situation. Desert plants will be more tolerant of low-moisture situations.

Bio Talk

Plants and Animals Compete for Resources

It is not uncommon for two organisms to compete with one another. Often, this happens when there is a limited supply of resources like water, food, sunlight, or space. If there are not enough resources to satisfy the needs of both organisms, they will compete with one another.

Sometimes this competition occurs between members of the same species. Male birds of the same species will battle each other fiercely for the ownership of a territory. The territory would allow sufficient food and habitat for the male and the female birds to make a nest, eggs, and feed offspring. Competition among individuals of the same species is a very important factor in evolutionary change.

Plants must deal with competition in different ways. Usually the plants that grow the tallest or establish the "best" root system are the survivors. In some cases, plants even secrete chemicals, which prevent the seeds from other plants from growing.

Competition may occur among members of different species. In this case, limited resources are usually the cause for competition.

Female eagles usually lay one to three eggs a few days apart. One to two days is a normal age difference between eaglets. Older hatchlings often dominate the younger ones for food. In a three-egg brood, the third chick has little chance of survival. A large number of adult eagles are non-breeders, probably due to competition within the species and variable annual food supply.

BioTalk

Teaching Tip

The text states that "Competition among individuals of the same species is a very important factor in evolutionary change." This concept is dealt with in greater depth in the next chapter. However, you may wish to direct students' attention to the statement at this time. Ask them to explain in their own words what the sentence means. Ask them to speculate how competition within a species can lead to evolutionary change.

A Vote for Ecology

Bio Words

nonnative (exotic, alien, introduced, or non-indigenous) species: any species, including its seeds, eggs, spores, or other biological material capable of propagating that species, that is not native to that ecosystem

invasive species: a nonnative species whose introduction does or is likely to cause economic or environmental harm or harm to human health

Purple loosestrife can produce up to 2.7 million seeds per plant yearly and spreads across approximately one million additional acres of wetlands each year.

Scientists have found that usually only one species survives in laboratory experiments studying this type of interaction. However, the real world is a much more complex situation. At times two competitive species can exist together. For example, both hawks and owls hunt mice. Hawks hunt by day and owls hunt by night.

Introducing Nonnative Species

Sometimes a new species that has not been there before is introduced into an ecosystem. This is called a **nonnative species.** (Other terms that you may hear used to refer to these organisms are **exotic**, **alien**, **introduced**, or **non-indigenous**.) The introduction can be either intentional or accidental. Introduction of a nonnative species can have negative effects on the ecosystem. When this happens the species is considered invasive. It can cause economic or environmental harm, or harm to human health. **Invasive species** can be plants, animals, and other organisms (e.g., microbes). All these terms are clearly defined by the U.S. government (Executive Order 13112).

Why is the government so concerned with invasive species? Why should the public be concerned? Most scientists believe that on a

Teaching Tip

In 1934 Soviet ecologist G. F. Gause completed a classic series of experiments to study the competition between two species. He used the single-celled *Paramecium*. He grew species in culture under two different conditions. Under one condition a species of *Paramecium* was grown alone in a culture containing a limited amount of bacterial food. Under the second condition the two species were grown together in the same vial. Whereas the population survived if grown separately, only one species survived when grown together with another. These and other similar experiments led ecologists to propose the principle of competitive exclusion. It states that two species cannot live together at the same time in the same place if they are limited to one or more limited resources. Stated in another way, no two species can occupy the same niche.

It is interesting to note that when Gause grew the two different species of *Paramecium* in the same tube, both species survived. One species fed on the bacteria at the bottom, and the other species fed on the bacteria suspended in solution.

global basis, next to habitat destruction, the second greatest destroyers of biodiversity are invasive species. In some cases, the result is the extinction of an entire existing species.

Species have sometimes invaded new habitats naturally. However, human actions are the main means of invasive species introductions. When people settle far from home, they often bring with them familiar animals and plants. Other species, like rats, make the trip unintentionally. In their new habitat there may have fewer predators or diseases, so their populations grow out of control. Organisms that they might normally prey on may not have evolved defense mechanisms. Native species may not be able to compete successfully for space or food and so are often pushed to extinction. The spread of nonnative organisms destroys healthy, diverse ecosystems. It replaces them with biologically impoverished, homogeneous landscapes.

It is unfortunate, but increased travel and trade are providing many new opportunities for the spread of nonnative species. In addition to which, one important feature that makes a community susceptible to invasion by nonnative species is the level of human-induced disturbance. For example, nonnative birds such as European starlings and house sparrows do well in ecologically disrupted areas such as cities, suburbs, and farms.

One hundred house sparrows were introduced into Brooklyn in the 1850s. From this initial introduction, the species expanded throughout the eastern United States and Canada. House sparrows are closely tied to human activity. This sparrow is usually absent from extensive woodlands and forests and from grasslands and deserts.

Teaching Tip

The government maintains extensive databases on invasive species that are easily accessible by the Internet. Some of these databases are regional in scope. Take some time to familiarize yourself with invasive species in your region.

A Vote for Ecology

Reflecting on the Activity and the Challenge

In this activity you looked at the competition among organisms. Sometimes this competition occurs among members of the same species. Other times it occurs among members of different species. Competition is not always bad. In fact, it plays an important role in evolutionary development. However, when a nonnative species is introduced into an ecosystem, the competition can have devastating results. Consider how your issue relates to the competition among organisms. Has the environmental issue that you are investigating for the **Chapter Challenge** been caused by competition?

Penguins breed in colonies and can be fiercely territorial. Since nearly all of Antarctica is covered in ice, competition for breeding space among penguins is great.

Biology to Go

1. Why does competition occur among organisms?
2. What would be an advantage to competition between organisms of the same species?
3. Why do new species that are introduced into an ecosystem often become invasive?
4. What type of evidence would you look for in a natural setting to indicate that there is competition taking place among the plants?
5. Europeans, and their descendants in North America, often describe humans as being at the center of change. Not only do humans cause environmental changes, they are also responsible for those changes. In this worldview, the ideal human acts as a steward or protector for an ecosystem. By contrast, First Nations peoples often describe humans as belonging to an ecosystem. In this worldview, the ideal human lives in harmony with the ecosystem. How would the two worldviews differ in describing what has happened to ecosystems in the United States over the last century?

Biology to Go

1. When resources are limited, competition occurs between organisms for these resources. Resources can include food, water, space, sunlight.

2. Competition between organisms of the same species tends to stabilize a population. It is also an important factor in evolutionary change. It ensures the survival of the fittest of a given species.

3. New species introduced into an ecosystem may lack any predators and therefore may reproduce at a very rapid rate. Also, the prey of the new species may not have developed any defense mechanisms against the introduced predator.

4. In a natural setting one would expect plants to be reasonably well spaced.

Inquiring Further

1. Investigating allelopathy

The production and release of substances by a plant that are toxic to neighboring plants is called allelopathy. Familiarize yourself with an allelopathic species in your area. Design an experiment to test which part of the plant is most toxic. Be sure to have your teacher approve your procedure if you plan on carrying out your experiment.

2. Invasive species

Research an invasive species in your area. Report your findings to the class.

Allelopathy benefited the sunflower growing in the wild. It reduced competition for nutrients, water, and sunlight. However, allelopathy works against sunflower crops. Sunflower crops must be rotated to avoid buildup of the "poison" in the soil.

Inquiring Further

1. **Investigating allelopathy**
   Let students know that allelochemicals can be found in leaves, roots, stems, flowers, and fruits of plant. Leaves are the major source in most species. The roots generally produce smaller quantities and the compounds are less toxic than those found in the leaves. They will need to know this in order to design an experiment.

2. **Invasive species**
   As mentioned earlier, the government maintains extensive databases on invasive species. These databases are easily accessible by the Internet. Some of these databases are regional in scope. Students can use these databases to find the names of invasive species in their area.

Assessment Opportunity

Whether students actually carry out the experiment, this **Inquiring Further** provides excellent practice in designing an experiment. It could provide an opportunity to assess if students understand the idea of variables and the need for controls in an experiment.

ACTIVITY 6– SUCCESSION IN COMMUNITIES

Background Information

Succession

Succession involves change that is not immediate but rather slow and gradual. Succession can either be primary or secondary. **Primary succession** occurs in an area where initially no community existed. For example, primary succession takes place when a glacier recedes and leaves behind barren land. It can also occur on a newly formed volcanic island. **Secondary succession** takes place when a community replaces an already existing community. For example, secondary succession takes place when a forest community is destroyed by fire. A community was already there before its destruction by the forest fire. For primary succession to take, rocks must first be broken down into soil. In secondary succession, soil is already present. This explains why it takes longer for primary succession to happen.

In the "lifeless" environment, the very first community to appear is the **pioneer community**. They are plants that lie close to the ground and can withstand high temperatures from full exposure to the Sun. Despite their short life, they make up for it by their fast dispersal of a lot of seeds. These short annual plants contribute organic material to the soil resulting in more favorable conditions present for taller plants that cannot tolerate full sunlight.

Replacements continue until a mature or a stable community called the climax community is established. A **climax community** is reached when no other community replaces it. Communities that are formed between the pioneer and the climax communities are referred to as **seral stages.**

A climax community is best exemplified by biomes like the desert with its characteristic flora and fauna, which are dictated by specific temperature and rainfall patterns.

This activity focuses on the observational skills of the students as they visually note the changes that occurred: nine months, 3 years, 13 years, 23 years, and 47 years after a volcano erupted. As they do their visual inspection, the students are encouraged to compare the events along the coastal areas and those in the inland areas. After their comparisons are made, the students are expected to explain why there is a difference and how the difference occurred.

Goals and Assessments

Goal	Location in Activity	Assessment Opportunity
Investigate succession after a natural disaster.	For You To Do	Student observations are similar to those listed in the Teacher's Edition. **Biology to Go** Question 1. **Inquiring Further** Question 3.
Distinguish between primary and secondary succession.	BioTalk	**Reflecting on the Activity and the Challenge** **Biology to Go** Questions 2–4. **Inquiring Further** Questions 1–3.
Explain how human activities can lead to succession.	BioTalk	**Reflecting on the Activity and the Challenge** **Biology to Go** Question 5.

Activity Overview

This activity provides an opportunity for the students to learn about succession. They do a visual scrutiny of a pictorial diagram on the changes that occurred along the coastal areas and in the inland areas after two volcanoes erupted on the island of Krakatoa. The main purpose of this activity is to hone the students' observational skills as they investigate the sequence of changes that take place in succession. The students are also given a chance to explain the difference in succession that they note on the two areas affected by the volcanic eruption. This activity further brings into focus the role that humans have in shaping their environment.

Preparation and Materials Needed

Preparation

Before this activity, remind the students to bring their books so that the required diagram analysis on the succession on Krakatoa will go smoothly. As a backup, you can have copies made of **Blackline Master Ecology 6.1: Succession on Krakatoa.** Although the analysis may be done in groups, ensure that each student has a copy of the diagram. To facilitate the learning process, you should have a deeper understanding of what the questions are asking for, to help the students be more focused in their study of the diagram.

Materials/Equipment Needed

There are no materials needed or equipments to be used in this activity.

Learning Strategies for Students with Limited English Proficiency

1. Point out new vocabulary in context. Practice using the words as much as possible.

volcanic island	tropical forest	succession	pioneer community
coastal	inland	annual	climax community
perennial	primary	secondary	temperature pattern
seral stages	biota	biomass	rainfall pattern
organic matter			

2. Discuss with the class the questions in **For You To Do**. Ask the students what the questions mean to them. This way, you can have the students fully focused with a clearer understanding of what they are supposed to target in their diagram analysis.

3. Discuss the difference between terms that qualify numbers like few, some, many, and abundant.

4. Have the students work in groups of four. Each student assumes a role from the four possible responsibilities – **reader**, **clarifier**, **questioner**, and **summarizer**. The **reader** reads what the book says and at the same time records what the group's answers are to the questions. The **clarifier** takes note of unfamiliar words that need to be made clear for the group to have a better understanding of the reading material. The **questioner** reminds the group of the questions that need to be answered like those found in **For You To Do.** This student also writes down any possible question that the group can think of. The **summarizer** synthesizes the concepts that the group was able to learn in the activity.

A Vote for Ecology

Activity 6 Succession in Communities

GOALS

In this activity you will:

- Investigate succession after a natural disaster.
- Distinguish between primary and secondary succession.
- Explain how human activities can lead to succession.

What Do You Think?

Following a forest fire, all that remains is a charred landscape. Yet, within a few weeks the ground begins to turn green as living organisms return.

- **From where does this new life come?**
- **How long will it take for the forest to return to its original condition?**

Write your answers to these questions in your *Active Biology* log. Be prepared to discuss your ideas with your small group and other members of your class.

What Do You Think?

- The new life comes from areas close to where the forest fire occurred.
  When plant seeds carried by the wind land on a charred landscape, they do not see it as charred but as a ground, rich in organic material provided by the ashes. With the good soil, the seeds germinate and grow.

- It will take decades for the forest to return to its original condition.
  The change is a very slow and gradual process.

Student Conceptions

Some students will find it hard to grasp change that occurs very slowly.
They understand change to be immediate and sudden.

Emphasize the point that succession is not just about one species succeeding another. Stress to the students that succession deals with a community replacing another one.

There may also be confusion with the difference between primary and secondary succession. If an existing community was there before the natural or human-caused disaster, then it is secondary succession.

For You To Do

This activity provides you with an opportunity to examine how "life re-establishes itself" after a devastating blow.

1. On August 27, 1883, two volcanoes located on a single island in the Indian Ocean erupted at the same time. The blast was so great that a hole about 250 m deep remained where the peak of the volcano had been. The eruption on the island of Krakatoa has been said to be the loudest noise ever heard on Earth. The blast was heard in Hawaii, several thousands of kilometers away. Hot cinders and lava covered the island. Before the eruption, Krakatoa had been covered with a tropical forest. The eruption completely destroyed life on Krakatoa and two other nearby islands.
2. Two months after the eruption, scientists visited the island of Krakatoa. They found it steaming from a recent rain that had fallen on the lava that was still hot. In some places, the volcanic ash was washing away. In other places the ash was still more than 60 m deep. No life was visible.
3. Scientists visited the island nine months after the explosion, and at later times, to record the living things on Krakatoa. Some of the data recorded is shown in the diagram on the next page. Look for some interesting patterns in the rebirth of life on the island of Krakatoa. Study the plant life. (Reports of the animal life are interesting but too limited to use.)

a) What happens to the number of kinds of plants as the years pass?

b) Is there a change in the number of kinds of plant life?

c) Do the numbers of some kinds of plants change more than the numbers of other kinds?

d) Where do you think these plants might have come from? What reason do you have for your belief?

e) How long a period was needed for the complete recovery of the forest growth?

f) Write a statement that will describe the kinds of changes that have taken place on the island since the eruption.

g) Compare the "rebirth" of plant life on the coastal areas with the rebirth of plant life in the inland areas. How would you explain the difference?

For You To Do

Teaching Suggestions and Sample Answers

3. a) **and b**) The number of kinds of plants increased as the years passed. In the inland areas, three years after the eruption, from a bare ground, 2 kinds of plants appeared – grasses and ferns. After 13 years, the ferns disappeared and 3 kinds of plants dominated – grasses, orchids and horsetail trees. After 23 years, the grasses, orchids and horsetail trees now shared the island with coconut trees, making a total of 4 kinds of plants. After 47 years, the kinds of plants remained the same and only the numbers increased.

   c) Yes. In the inland areas, in 10 years – from the 3rd year to the 13th year while the number of grasses increased, the ferns disappeared and orchids and horsetail trees appeared. Along the coastal areas, while the grasses, ferns and many tropical seashore plants disappeared; young coconut trees horsetail trees and sugar cane plants completely covered the area.

   d) The plants came from neighboring islands. Plants propagate through very efficient seed dispersal. One way of seed dispersal is the use of wind for the spread of the plant seeds. Another way that is useful on this island, is the water dispersal of seeds.

   e) Almost five decades was needed for the complete recovery of the forest growth.

   f) It takes decades for a forest to recover from a volcanic eruption. Plants that appear initially may either increase their numbers or totally disappear. Shorter plants are replaced by taller trees.

   g) Although the "rebirth" of plant life on the coastal areas is parallel to that in the inland areas, change occurred faster on the coastal areas. Comparing the two areas shows that change on the coastal area was a step ahead of the inland area. It took 13 years after the eruption, for the inland areas to be completely covered with young coconut trees, horsetail trees, and sugar-cane plants and 23 years for the inland areas to go through the same process. It appears that the coastal area's recovery happened sooner than that of the inland area.

   The coastal areas with their proximity to a large body of water had their temperature go down sooner than those in the inland areas. With the lower temperature and the presence of wind and water dispersed seeds, plants appeared sooner on the coastal areas compared with those in the inland areas.

Teaching Tips

Students who require extra individual attention may wish to use a copy of a table found in **Blackline Master Ecology 6.2: Comparison of Coastal and Inland Succession on Krakatoa** to fill in. Ask the students to identify the kinds of plants as well as their numbers in terms of few, some, many, and abundant.

NOTES

A Vote for Ecology

SUCCESSION ON KRAKATOA

COASTAL AREAS		INLAND AREAS
 Only algae and one lone spider found... mostly bare lava.	**3/4 years since eruption**	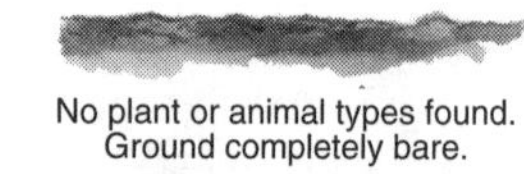 No plant or animal types found. Ground completely bare.
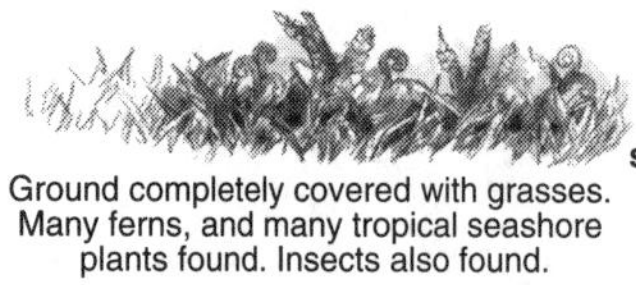 Ground completely covered with grasses. Many ferns, and many tropical seashore plants found. Insects also found.	**3 years since eruption**	A few grasses, many ferns and insects found.
 Completely covered with young coconut trees, horsetail trees, and sugar cane plants. Lizards as well as insects found.	**13 years since eruption** 	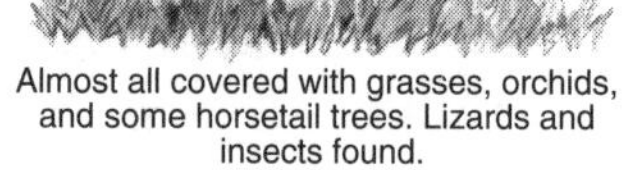 Almost all covered with grasses, orchids, and some horsetail trees. Lizards and insects found.
 Completely covered as before, but with a greater number of trees.	**23 years since eruption**	 Completely covered now with grasses, orchids, and groves of horsetail and young coconut trees.
 By now a dense forest covers the area. All the previously listed plants and animals are found in abundance.	**47 years since eruption**	 Inland areas now support same amount of plants and animals as the coast.

NOTES

BioTalk

Succession

The destruction of a mature forest by a severe fire is a devastating scene. Yet, even this charred scene holds promise of new life. Within a few weeks the ground will slowly turn green as annual and perennial plants return. These plants can tolerate full sunlight and the resulting high soil temperatures. They take root, grow, and reproduce in a soil made fertile by the mineral content of the ash. Within two or three years shrubs and young trees are evident and growing rapidly.

Forest fires are one of the most destructive natural forces known. While sometimes caused by lightning, nine out of ten forest fires are caused by humans. Natural-occurring fires are vital in maintaining healthy ecosystems. However, human-caused fires have devastating effects on both wildlife and human lives.

A few years later, an untrained observer would probably never know that the area had once been burned out. Over the long term, the forest will again reach maturity. This pattern is not limited to forests. It occurs in many other environments. This process of re-growth follows an environmental change. It is called **succession**. It describes the gradual change in an area. The change takes place as the area develops toward a final stable community. In every case, the final community that can exist is determined by the abiotic factors of the area.

Bio Words

succession: the slow and orderly replacement of community replacement, one following the other

BioTalk

Teaching Tip

Ask the students if all forest fires are bad. Students may not know that some forest plants require intense heat from fires for their seeds to germinate.
It is only through forest fires that plants will get their much-needed nutrients from the ashes that are produced. It also ensures them enough sunlight since the fire destroys the plants that block the Sun.

A Vote for Ecology

Bio Words

primary succession: the occupation by plant life of an area previously not covered with vegetation

secondary succession: the occupation by plant life of an area that was previously covered with vegetation and still has soil

There are two types of succession: primary and secondary succession. **Primary succession** occurs in an area where no other community existed before. For example, this could happen on land left behind by a receding glacier. It could also happen on a newly formed volcanic island. **Secondary succession** occurs following destruction of a community. The re-growth after a forest fire is an example of secondary succession. Since soil is already present, the long time needed for soil to form in primary succession is not necessary.

Primary succession occurs on rock left behind by retreating glaciers, and transforms it into a living community. The process must begin with organisms that form organic soil, the pioneers or soil builders. This soil will be necessary to provide for the next group of plants to succeed.

After a forest fire, a sequence of ecological responses begins. Amid the charred forest remains, a pioneer community is established.

Teaching Tip

You can discuss with the students how rocks are broken down into soil. Rocks are broken down by lichens and plants that release acids that eventually dissolve the rock minerals. With the production of carbon dioxide and water during cellular respiration, carbonic acid is produced.

Each community goes through a succession of plant and associated animal species. The first community to appear is the **pioneer community**. It includes plants that are able to tolerate sunlight and the resulting high temperatures. This development of vegetation sets up new ground-level conditions. Eventually, conditions become more favorable to other plants that cannot tolerate full sunlight. These plants tend to be taller than the pioneer plants. This process continues through several in-between communities called **seral stages**. The plants and animals and their wastes at each stage contribute to the community development until a final community is reached. This final community is the one that can continue to perpetuate itself. It is called the **climax community**. Generally, both the biomass and the nonliving organic matter increase during the stages of succession. They then level off when the climax community is reached. The climate plant and animal life (biota) vary greatly by area. The types depend upon the temperature and rainfall patterns.

Bio Words

pioneer community: the first species to appear during succession

seral stages: the communities in between the pioneer and climax community during the stages of succession

climax community: the final, quite stable community reached during the stages of succession

Reflecting on the Activity and the Challenge

In this activity you observed some of the changes that occur after a dramatic environmental change. Sometimes, human activities are responsible for the environmental changes. Consider how the issue that you are researching for the **Chapter Challenge** came about. Can secondary succession be part of the solution to the problem?

Is the succession that occurs after a volcanic eruption primary or secondary succession?

Teaching Tip

Secondary succession occurs after a volcanic eruption. A community was already in existence before the eruption. Soil is still present underneath the ashes.

Biology to Go

1. What is succession?
2. Explain the difference between primary and secondary succession.
3. Which community would support the greatest number and diversity of organisms, the pioneer community or the intermediate stages? Explain your answer.
4. Explain how abiotic factors change within a community as a result of the succession of vegetation.
5. Give two examples of how human activities can lead to secondary succession.

Inquiring Further

1. Primary succession

How does life establish itself on a rock surface? Describe the stages of succession that must occur to "transform" rock into a climax community.

2. Hydrarch succession

Hydrarch succession is the name for primary succession in a new freshwater environment. What type of organisms constitute the pioneer, seral, and climax communities in the stages of succession in fresh water?

Volcanoes provide a unique opportunity to study plant succession, animal behavior, evolutionary and geologic processes, and ecology. Understanding how natural systems respond to disturbances is essential in facing environmental challenges of the future.

3. Succession and Mt. St. Helens

At 8:32 Sunday morning, May 18, 1980, Mt. St. Helens in southwestern Washington state erupted. About 600 km^2 of forest was blown over or left dead and standing. The eruption lasted nine hours, but Mt. St. Helens and the surrounding landscape were dramatically changed within moments. Scientists and visitors follow the changes in the landscape and the volcano. Surviving plants and animals rise out of the ash, colonizing plants catch hold of the earth, birds and animals find a niche in a different forest on the slopes of Mt. St. Helens. Research the succession pattern on Mt. St. Helens.

Biology to Go

1. Succession is the slow and orderly replacement of a community.
2. The difference between primary succession and secondary succession lies in the existence or non-existence of vegetation cover before the disaster occurred. If the area was not previously covered with vegetation, then it is primary succession. If the area was previously covered with vegetation, then it is secondary succession.
3. The intermediate stages would support a greater number and diversity of organisms than the pioneer community. Between the pioneer community and the climax community are seral stages or intermediate stages. As the pioneer stage gives way to a seral stage and to another seral stage, more complex relationships like competition and predation are created. New niches are formed resulting in more species diversity.
4. Plants transpire and in the process lose water into the atmosphere. They take in carbon dioxide that is used for photosynthesis. They also absorb radiation from the Sun. These result in a moderate temperature and help maintain a stable climate.
5. Human-caused fires destroy thousands of forest acres. One solution to this disaster is secondary succession. Deforestation for agricultural purposes can also be solved by secondary succession.

Inquiring Further

1. **Primary succession**
   A rock can be broken by lichens and plants that produce acids. The acid dissolves the rock minerals. After the rock is transformed into soil by the pioneers or soil builders, the next groups of plants that are usually taller succeed. Several seral stages occur that are eventually followed by the climax community.

2. **Hydrach succession**
   The pioneer community in hydrarch succession are submerged aquatic plants. Seral communities may consist of cattails and bulrushes, which take root in shallow water's mud. Willow thickets found along the banks constitute another seral community. The climax community, eventually, is a mature forest that may be in the form of a conifer forest.

3. **Succession and Mt. St. Helens**
   Two groups of scientists made predictions as to the recovery of Mt. St. Helens. One group predicted that pioneer plants would start and that succession would end in a mature forest. The other group thought that succession would occur by chance where a seed borne by the wind would take root and influence the kinds of plants that would grow next. This is in contrast to the earlier thought of an orderly process. Recovery is occurring as predicted by the two groups. A typical colonizer started as one of the first plants. When this colonizer died, its stalk and leaves collected particles of sand and dust. This allowed the perennials to grow resulting in an orderly succession. In the other barren areas, trees, which were usually alder, grew. Alders can survive even in poor soil that is deficient in nutrients because of its symbiotic relationship with nitrogen-fixing bacteria.

Teaching Tip

You may show a video on the Mt. St. Helens eruption and its recovery.

NOTES

Blackline Master Ecology 6.2: Comparison of Coastal and Inland Succession on Krakatoa

Number of Years Since Eruption	Coastal Areas	Organisms Found	Inland Areas

Answer Key
Blackline Master Ecology 6.2: Comparison of Coastal and Inland Succession on Krakatoa

Number of Years Since Eruption	Coastal Areas	Organisms Found	Inland Areas
3/4	yes	algae	none
	1	spider	none
3	abundant	grass	few
	many	ferns	many
	many	tropical seashore plants	none
	yes	insects	yes
13	abundant	young coconut trees	none
	abundant	horsetail trees	some
	abundant	sugar cane plants	none
	abundant	grasses	abundant
	none	orchids	abundant
	yes	lizards	yes
	yes	insects	yes
23	more abundant	young coconut trees	many
	more abundant	horsetail trees	many
	abundant	sugar cane plants	none
	abundant	grasses	more abundant
	none	orchids	more abundant
	yes	lizards	yes
	yes	insects	yes
47	abundant	young coconut trees	abundant
	abundant	horsetail trees	abundant
	abundant	sugar cane plants	abundant
	abundant	grasses	abundant
	none	orchids	abundant
	abundant	lizards	abundant

ACTIVITY 7– THE WATER CYCLE

Background Information

The Water Cycle

Where there is water, there is life. Living things on Earth have a connection with water and water likewise is linked with them. Plants absorb water through their roots and they transpire and lose water into the atmosphere. Animals get their water by drinking it or by processing their food. Living things perform cellular respiration and produce water as a byproduct. This water is also released into the atmosphere. Another source is the evaporation of water from other bodies of water. All of these contribute to the water vapor in the atmosphere. With cooler temperature, the water vapor goes through the process of condensation. This is followed by precipitation that allows the water, either in liquid form as rain, or solid form as snow or hail, to return to Earth.

Transpiration

Plants are rooted where they are and the only way they can get food is by making it themselves. They do this by utilizing the Sun's energy in the process of photosynthesis. Energy that the plants need in order to maintain their daily functions come from the very food that they synthesized. This process of getting energy is cellular respiration. A byproduct produced in cellular respiration is water. As the plants take in carbon dioxide through their leaf openings called stomata, water in the form of water vapor leaks out. When leaves lose water into the atmosphere, the process is called transpiration.

This activity focuses on the skills involved in doing an experiment. The students, as they set up and measure transpiration rates, are given the opportunity to figure out the factors that could affect this rate. As they do their experiment, the students are encouraged to look for possible sources of experimental errors.

Goals and Assessments

Goal	Location in Activity	Assessment Opportunity
Measure the amount of water transpired by a plant.	**For You To Do**	**For You To Do** Steps 8 (a)–(c) **Biology to Go** Questions 1 and 5
Describe the processes that take place in the water cycle.	**BioTalk**	**Biology to Go** Question 1
Provide examples of how human activities can affect the water cycle.	**BioTalk**	**Biology to Go** Questions 3, 4, and 5
Model the effects of acid rain on an ecosystem.	**Inquiring Further**	**Inquiring Further** Ecocolumn model

Activity Overview

This activity provides an opportunity for the students to learn about the water cycle and also the role of transpiration in the cycle. They conduct an experiment measuring the rate of transpiration. The main purpose of this activity is to hone the students' skills in doing experiments as they investigate factors that could affect experimental results as well as possible experimental errors. The students are also given a chance to investigate and appreciate human's role on the water cycle.

Preparation and Materials Needed

Preparation

At least one day before the experiment, have all the materials ready and place them on a table accessible to all the students. Have **For You To Do** as an assigned reading the night before the experiment. This is to familiarize the students with what is expected of them while doing the experiment the next day. Have the groups count off from 1 to 4. Give designated functions for each group member: 1 is responsible for getting the materials, 2 is the reader, 3 is the recorder, and 4 is responsible for cleaning up the assigned work area.

On the day of the experiment, list the required materials for each group on the board. Write the assigned function of each group member.

Materials/Equipment Needed (per group)

- 40-cm piece of plastic tubing
- 0.1 mL pipette
- water tank or wide-mouthed jar made of plastic or glass
- branch of Coleus or Zebrina
- garden scissors
- paper towels
- iron clamp
- timer
- graph paper
- ruler
- calculator

Learning Strategies for Students with Limited English Proficiency

1. Point out new vocabulary in context. Practice using the words as much as possible.

evaporation	condensation	precipitation	water (hydrologic) cycle
sublimation	groundwater	infiltration	aquifer
solid	liquid	submerge	water vapor
snow	glacier	runoff	plot
gas			

2. For homework – have the students draw the steps in the experiment. This pictorial representation will help them in knowing what to do during the experiment itself.

3. Have the students work in groups of four. Each student assumes a role from the four possible responsibilities – **reader**, **clarifier**, **questioner**, and **summarizer**. The **reader** reads what the book says and at the same time records what the group's answers are to the questions. The **clarifier** takes note of unfamiliar words that need to be made clear for the group to have a better understanding of the reading material. The **questioner** reminds the group of the questions that need to be answered like those found in **For You To Do**. This student also writes down any possible question that the group can think of. The **summarizer** synthesizes the concepts that the group was able to learn in the activity. Allow the summarizer to draw it if he/she prefers to do so.

4. Labeling **Blackline Masters 7.1 and 7.3** both in English and in their native language will provide students with additional vocabulary practice.

Activity 7 The Water Cycle

GOALS

In this activity you will:

- Measure the amount of water transpired by a plant.
- Describe the processes that take place in the water cycle.
- Provide examples of how human activities can affect the water cycle.
- Model the effects of acid rain on an ecosystem.

What Do You Think?

According to William Shakespeare, Caesar used part of his last breath to utter the words "Et tu, Brute" (even you, Brutus). His last breath would have been partly made up of water.

- **Is it possible that the molecules of water that Caesar exhaled many centuries ago, are still a part of today's environment?**
- **Is it possible that these molecules could become a part of you?**

Write your answer to these questions in your *Active Biology* log. Be prepared to discuss your ideas with your small group and other members of your class.

For You To Do

Plants absorb water through their roots and return water to the atmosphere through the process of respiration. In this activity you will measure the amount of water transpired by a plant over a period of time.

What Do You Think?

- Yes. Since water is supposed to be cycled, it is possible that the water molecules exhaled by Caesar centuries ago are still part of today's environment.
- Yes. Since the same water molecules exhaled by Caesar centuries ago are cycled, they can become a part of us when we drink water. Another way that they can become a part of us is through the food that we eat.

Student Conceptions

Students may not know that plants transpire and lose water through their leaves. Animals have their noses to breathe out of while plants have their leaves, more specifically their stomata, which are openings on the underside of the leaves.

A Vote for Ecology

1. You will first construct a very simple "meter" to measure the loss of water. Place the tip of a 0.1 mL pipette into a piece of plastic tubing about 40 cm long.
2. Submerge the tubing and the attached pipette under water in the sink or a tray. Fill both with water. Make sure that all the air is drawn out of the tube and pipette. Leave the assembly under water while doing the next step.
3. Choose a branch from a plant. Suggested plants include Coleus and Zebrina. Submerge the end of the branch under water and make a small, slanted cut. This step is very important to ensure that no air bubbles are introduced into the xylem cells and the water will flow easily. Do not get the leaves wet. If you do, dry them gently with a paper towel before you begin your experiment.
4. While the branch and tubing are still under water, push the freshly trimmed end of the branch into the open end of the plastic tube. The end of the branch should be about 1.5—2.0 cm in the tube. There should be a very tight fit between the stem of the branch and the tube.

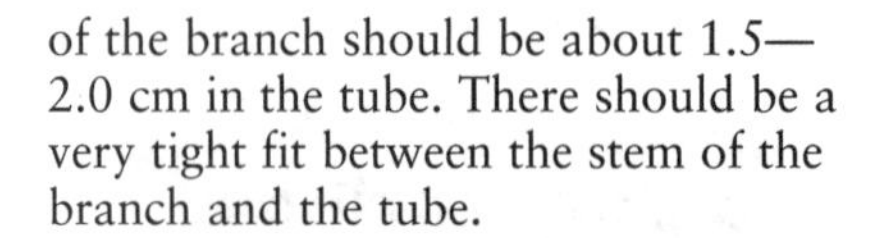

Be careful when cutting the plant. Cut away from yourself. Report any injuries.

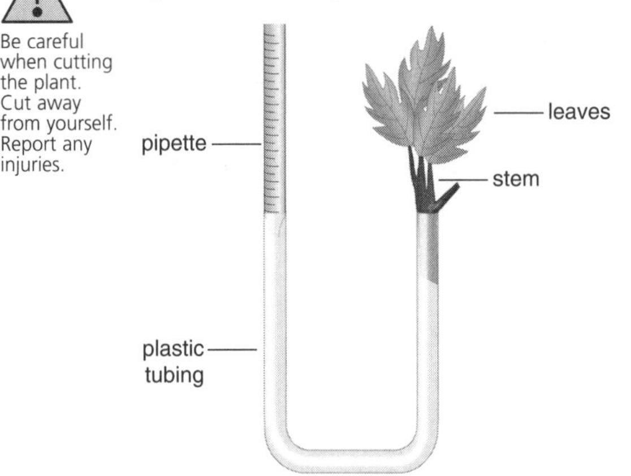

5. Bend the tubing into a U shape, as shown in the diagram. Clamp the tubing with the branch and the pipette onto a ringstand.
6. Once the "meter" is set up, wait about five minutes to make sure the plant is transpiring. After this initial waiting period, read the water level at "0 minutes." Then read the water level in the pipette every three minutes for a total of 30 minutes.
   a) Record your observations in a table.
7. At the end of your experiment, cut the leaves off the plant stem. Find the area of the leaf surfaces. You can do this by arranging all the cut-off leaves on a grid. Trace the edge of the leaves on to the grid. Count all of the grids that are completely within the tracing and estimate the number of grids that lie partially within the tracing.
   a) Record the area of the leaves in square centimeters.
   b) Calculate the water loss per square centimeter of leaf surface. Divide the water loss at each reading by the leaf surface area you calculated.
   c) Calculate and record the average loss per square centimeter for the class.
   d) Graph the loss per square centimeter over time.

Wash your hands after the activity.

For You To Do
Teaching Suggestions and Sample Answers

6.–7. Students' answers will vary, depending on the size of the leaf used.

8. a) The factors that might affect the results of this experiment and their effects are:
   - The number of leaves in the branch. The more leaves there are, the more water is lost through transpiration.
   - The kind of plant used. Some plants lose more water through their leaves compared to other plants.
   - The "freshness" of the plant. The fresher the plant, the more transpiration occurs.
   - The humidity in the classroom. The lower the humidity, the more water is lost by transpiration.
   - The temperature in the classroom. The higher the temperature, the more water is lost by the leaves.
   - The presence of a light source in the classroom. The more intense the light is, the more water is lost through transpiration.
   - The "fit" of the branch on the plastic tubing. A "fit" that is not airtight will result in a lower amount of transpired water.
   - The "fit" of the pipette on the plastic tubing. A "fit" that is not airtight will result in a lower amount of transpired water.

   b) To reduce errors:
   - Have an airtight "fit" between the plastic tubing and the pipette/branch; a paraffin wax can be used.
   - Only fresh branches should be used.
   - The experiment should be done near a window to have a light source or close to an artificial light source like a fluorescent bulb.

   c) Investigations designed by the students will vary:
   - If the factor is light, the hypothesis may state that if the plant is exposed to light then the rate of transpiration is high. The control will have light and the experimental side will not.
   - If the factor is temperature, the hypothesis may state that if the plant is exposed to high temperature, then there will be a higher rate of transpiration. The control will have normal room temperature and the experimental side will have higher or lower temperature.

Teaching Tips

Using **Blackline Master Ecology 7.4: Transpiration Experiment Table,** discuss with the students the difference between the actual reading in the pipette and the adjusted reading. Discuss the units of measurement to be used.

Using **Blackline Master Ecology 7.5: Area of the Leaf Surfaces,** point out to the students that they have to make approximations of those squares that are not fully within the tracing. Discuss the units of measurement to be used.

Using **Blackline Master Ecology 7.6: Rate of Transpiration,** discuss that this graph shows the amount of water lost per square centimeter over time. Have the students figure out what they will put/plot on the *x*-axis and the *y*-axis. You can explain that the indicator of water loss is put/plotted on the *y*-axis. The indicator is also known as the dependent variable. The time that elapsed independent variable is on the *x*-axis and the dependent variable is plotted on the *y*-axis.

NOTES

e) Assuming that your plant stem continued transpiration at the same rate, estimate the total volume of water that might be transpired in 24 h.

8. Use the results of your experiment to answer the following questions.
   a) List some of the factors that might affect the results of this experiment. Explain how each factor might affect your results.
   b) How could you improve the design of the experiment to reduce errors?
   c) Consider what factors could affect the rate of transpiration of a plant. Develop a hypothesis and design an investigation to test your hypothesis.

The Water Cycle

Water is necessary to life in many ways. Land plants absorb water from the soil and land animals drink water or obtain it from their food. Water constantly bathes organisms that live in ponds, lakes, rivers, and the oceans. Other organisms rely on water to carry nutrients to their cells and organs. The cytoplasm in cells is mainly water.

Every day about 1200 km^3 of water evaporates from the ocean, land, plants, and ice caps. An equal amount of precipitation falls back on the Earth.

BioTalk

Teaching Tip

Ask what is happening in the photograph. Ask what the unit km^3 means. Explain that this is a measurement of volume *(V)* that is a three-dimensional measurement involving length *(l)*, width *(w)* and height *(h)* where $V = l \times w \times h$.

A Vote for Ecology

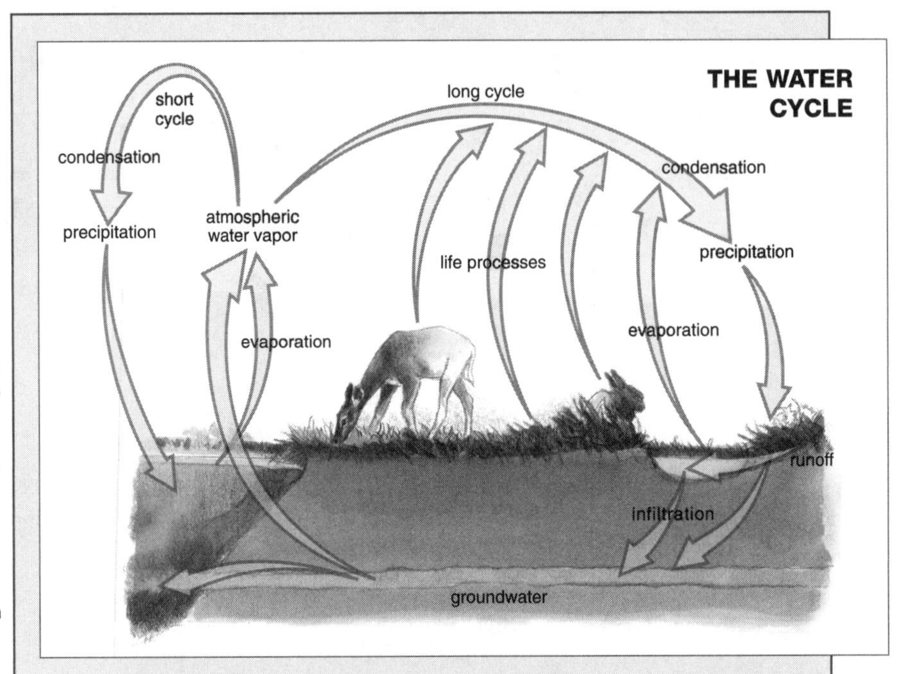

The volume of water in the biosphere remains fairly constant through time. In fact, the water that you used today has been around for hundreds of millions of years. It has probably existed on the Earth's surface as a liquid, a solid, and a vapor. However, water is always moving from place to place. It is forever changing from one state to another. This complicated movement of the Earth's water is called the **water cycle** or **hydrologic cycle**. Some of the pathways of this cycle are shown in the diagram above.

One of the largest reserves of water on Earth is found in the oceans. The oceans contain about 97% of the Earth's water. Other surface water includes lakes, rivers, estuaries, marshes, and swamps. By contrast, the atmosphere holds less than 0.001% of the Earth's water. This means that rapid recycling of water must take place between the Earth's surface and the atmosphere.

By absorbing heat energy from the Sun, some of the water on the Earth's surface changes to water vapor by **evaporation**. It rises upward into the atmosphere until it reaches a point where the

Bio Words

water (hydrologic) cycle: the cycle or network of pathways taken by water in all three of its forms (solid, liquid, and vapor) among the various places where is it temporarily stored on, below, and above the Earth's surface

evaporation: the process of changing from a liquid to a gas

condensation: the process of changing from a gas to a liquid

precipitation: water that falls to the Earth's surface from the atmosphere as liquid or solid material in the form of rain, snow, hail, or sleet

runoff: the part of the precipitation appearing in surface streams

groundwater: water contained in pore spaces in sediments and rocks beneath the Earth's surface

infiltration: the movement of water through pores or small openings into the soil and porous rock

aquifer: any body of sediment or rock that has sufficient size and sufficiently high porosity and permeability to provide an adequate supply of water from wells

Teaching Tips

Using **Blackline Master 7.1**, discuss with the students the changes of state. You may wish to provide the students with a copy of the diagram so that they can label the changes themselves.

Blackline Master 7.3 is provided so that students can label some of the pathways of the water cycle.

Assessment Opportunity

You may provide your students with a quiz to assess how well they understood the water-cycle diagram. The following questions may be included:

1. Name the two cycles involved in the water cycle and differentiate one from the other.

   (The short cycle involves the oceans/seas, while the long cycle includes life processes and groundwater; the short cycle occurs faster, while the long cycle takes longer to complete; with the short cycle involving the large bodies of water, more evaporation occurs and it occurs faster resulting in more water vapor formation that eventually produces more precipitation after condensation, while in the long cycle, the life processes involved take longer to add water vapor to the atmosphere, thereby taking longer to form precipitation after condensation.)
2. Explain the life processes involved in the long cycle.

   (Cellular respiration that occurs both in plants and animals releases water; transpiration that occurs in plants releases water through the leaves)
3. What are the similarities between the short cycle and the long cycle?

   (Both involve the processes of evaporation, condensation and precipitation)
4. What is infiltration and how does it contribute to the water cycle?

   (Infiltration occurs when water on the surface of the ground trickles down. It contributes to the water cycle when it joins the groundwater that eventually ends up in the oceans/seas.)

temperatures are low enough for the water vapor to condense to form tiny droplets of liquid water. This process is called **condensation**. These droplets of water are light. They collect around dust particles forming clouds or fog. They remain suspended in the atmosphere as clouds or fog and are supported by rising air currents and winds. When conditions are right, the droplets come together to form larger drops or sometimes ice crystals. Once the mass of the droplet or ice crystal can no longer be supported by air currents, **precipitation** occurs. Precipitation may take the form of rain, hail, sleet, or snow.

Snow falling high in the mountains or in the polar regions of the Earth may stay frozen there for years. Gradually, as layers of snow accumulate, the bottom layers of snow turn to ice, forming glaciers. Sometimes the snow or ice at the surface of the Earth can change directly back into water vapor. This process is called sublimation.

Approximately 1.7% of the water on Earth is stored in the polar icecaps, glaciers, and permanent snow.

Other precipitation lands on the surface of the Earth and flows along the surface as **runoff**. The ground runoff gathers in streams, lakes, and oceans, and the cycle then repeats itself.

However, some of the precipitation seeps into the Earth to form **groundwater**. This process is called **infiltration**. Sometimes the rock under the surface is very permeable. That is, water flows easily through it. In this case, some of the groundwater may seep to the surface, forming individual springs. **Aquifers** are large accumulations of underground water. They can provide an excellent source of water from wells. Groundwater

Teaching Tip

Using **Blackline Master 7.7: Major Biomes of the Earth**, discuss with the students where the ice caps, glaciers, and permanent snow are located. In your discussion, ask them how the global warming would affect the ice caps, glaciers, and permanent snow. Ask them further what that would mean in terms of sea level. *(With global warming, glaciers, icecaps, and permanent snow would very slowly melt and would then make the sea level higher.)*

Bio Words

transpiration: the emission of water vapor from pores of plants as part of their life processes

flow, although measurable, is much slower than the flow in streams and rivers. That is because the passageways through the pore spaces in the materials beneath the Earth's surface are very small. Nonetheless, regardless of its speed, groundwater eventually also returns to the rivers, lakes, and oceans. And, the water cycle continues.

Plants and animals also play a very important role in the water cycle. Water enters living organisms by osmosis. However, through cellular respiration, water is released back into the atmosphere. As you saw in this activity, plants, especially broadleaf trees and shrubs play a major role in the water cycle through the process of **transpiration**. Transpiration is the loss of water through the leaves of a plant.

The Human Impact on the Water Cycle

The Earth's water supply remains constant, but humans can interfere with the water cycle. As population increases, living standards rise, and the industry and economy grow, humans place a greater demand on the supply of freshwater. The amount of freshwater needed increases dramatically, yet the supply of freshwater remains the same. As more water is withdrawn from rivers, lakes, and aquifers, local resources and future water supplies are threatened.

A person can probably exist on about 4 L (four liters is about one gallon) of water a day for drinking, cooking, and washing. At present in the United States, people use almost 6000 L a day for their needs and comforts. These include recreation, cooling, food production, and industrial supply.

A larger population and more industry also mean that

Teaching Tip

Using **Blackline Master 7.8: Human Population Growth,** discuss the effect of the growth of human population on the freshwater resource of the Earth. Emphasize that the freshwater resource remains constant while the human population does not.

Also discuss how pollution from the power plants affects the water cycle. Sulfur dioxides (SO_2) are produced when fossil fuels are oxidized. Nitrogen oxides (NO_x) are produced when atmospheric nitrogen is oxidized in car engines and furnaces. SO_2 and NO_x when combined with the atmosphere's water vapor will form sulfuric acid (H_2SO_4) and nitric acid (HNO_3).

more wastewater is discharged. Domestic, agricultural, and industrial wastes include the use of pesticides, herbicides, and fertilizers. They can overload water supplies with hazardous chemicals and bacteria. Poor irrigation practices raise soil salinity and evaporation rates. Urbanization of forested areas results in increased drainage of an area as road drains, sewer systems, and paved land replace natural drainage patterns. All these factors put increased pressure on the water equation.

Pollutants that are discharged into the air can also affect the water cycle. Sulfur and nitrous oxides from the burning of fossil fuels, combustion in automobiles, and processing of nitrogen fertilizers enter the atmosphere. They combine with water droplets in the air to form acids. They then return to the surface of the Earth through the water cycle as acid precipitation.

Reflecting on the Activity and the Challenge

In this activity you observed one of the processes that take place in the water cycle. You learned that a great amount of water is transpired by a living plant. You also read about some of the other processes that are involved in the water cycle. The water cycle is very complex, and at any stage humans can have a significant impact. Perhaps the environmental issue you have chosen involves one part of the water cycle.

Biology to Go

1. Name and describe at least four processes that take place in the water cycle.
2. What is the energy source that drives the water cycle?
3. How has the water cycle determined partly where people live in the United States?
4. What would happen to the planet if the hydrologic cycle stopped functioning?
5. Describe three ways in which humans can have a negative effect on the water cycle.

Biology to Go

1. The processes in the water cycle and their description are:
   - Evaporation occurs when liquid water is turned into water vapor.
   - Condensation occurs when water vapor is turned into liquid with very low temperature.
   - Precipitation occurs when water from the atmosphere returns to the Earth's surface as either liquid or solid in the form of rain, snow, hail, or sleet.
   - Life processes involve the living organisms that contribute water into the atmosphere through cellular respiration and transpiration.

2. The Sun is the energy source that drives the water cycle.

3. People in the United States live where there is water. This excludes the desert because of the scarcity of water. With modernization, however, water is made available even in the desert areas like Las Vegas. Preference for areas where there are large bodies of water is also seen. These areas provide milder climates and temperatures that are not extreme.

4. If the water cycle or hydrologic cycle stopped functioning, then life would cease to exist. With the loss of 1200 km^3 of water each day from the ocean, land plants, and ice caps, water would eventually disappear and would never get to be replenished. With no formation of precipitation, drought would become common and death to all living things would result.

5. Ways in which humans can have a negative effect on the water cycle are:
   - With increasing human population, there is more demand on the freshwater resource, which remains constant. The big demand caused by the increasing population will result in very scarce freshwater resource.
   - With increasing population, an increase in the number of industries results, which produces more domestic, agricultural, and industrial wastes. These can overload water supplies with hazardous chemicals and bacteria.
   - With the clearing of forested areas for human purposes, natural drainage patterns are replaced by road drains, sewer systems, and paved land. These factors put increased pressure on the water cycle.
   - Pollutants discharged into the air from fossil fuel burning, automobile combustion, and nitrogen fertilizer processing, release sulfur and nitrous oxides. These oxides enter the atmosphere where they combine with water droplets and form acids. They return to the Earth in the form of acid precipitation.

A Vote for Ecology

Inquiring Further

Environmental models

Environmental models allow scientists to study what could happen to the plants and animals in an area if changes occurred. Models help check predictions without disrupting a large area.

Build an ecocolumn to research how acid rain affects an ecosystem. (You will be allowed to use household vinegar as the acid.) An ecocolumn is an ecological model that is especially designed to cycle nutrients.

Record the procedure you will use. Have your teacher approve your procedure before you create your model.

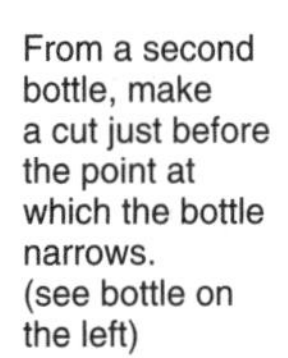

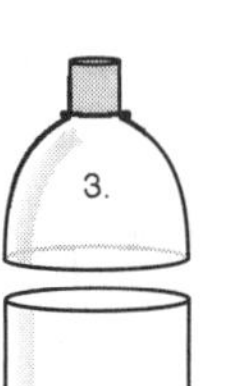

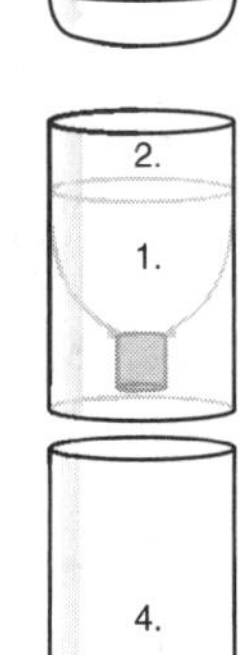

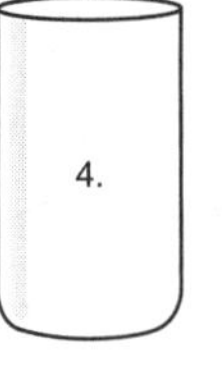

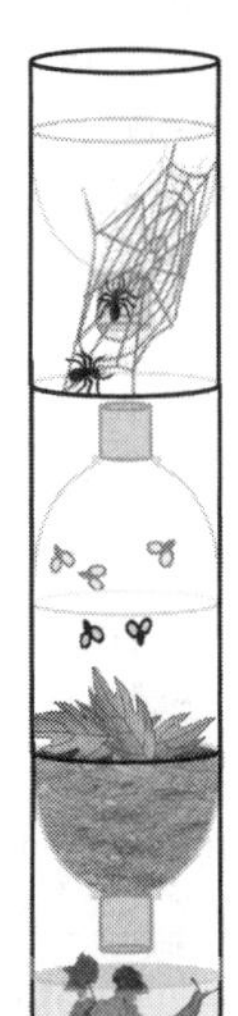

Inquiring Further

Environmental models

Students' procedures on creating an environmental model to research how acid rain affects an ecosystem will vary. The variation will be on the kinds of organisms that the students will use; the amount of water and vinegar; and the complexity of the ecocolumn to be built.

Teaching Tips

Setting up an ecocolumn is a very important modeling activity. We strongly recommend that this **Inquiring Further** be done by the class.

A month before this activity, you can start telling the students to bring 2-liter soda bottles. This will give you enough time to collect as many bottles as you can.

Discuss with the students that they have to figure out what would indicate the effect of the acid rain. Will this be the death of the organisms? Will this be the wilting of the plants? Tell them that these indicators are the dependent variables. Explain to them that the acid rain is what is causing the change that is shown by the indicators. The acid rain is the independent variable.

The students can make "windows" on the side of the plastic bottle. They can be holes through which acidity indicators like pH papers can be put through. The caps on the bottles can also be pierced with holes to allow fluids to be introduced into the ecocolumn.

A possible source of sulfur oxide could be the smoke after blowing out a burning matchstick. This could also be introduced into the ecocolumn through the "window."

Blackline Master Ecology 7.1: Changes of State of Water

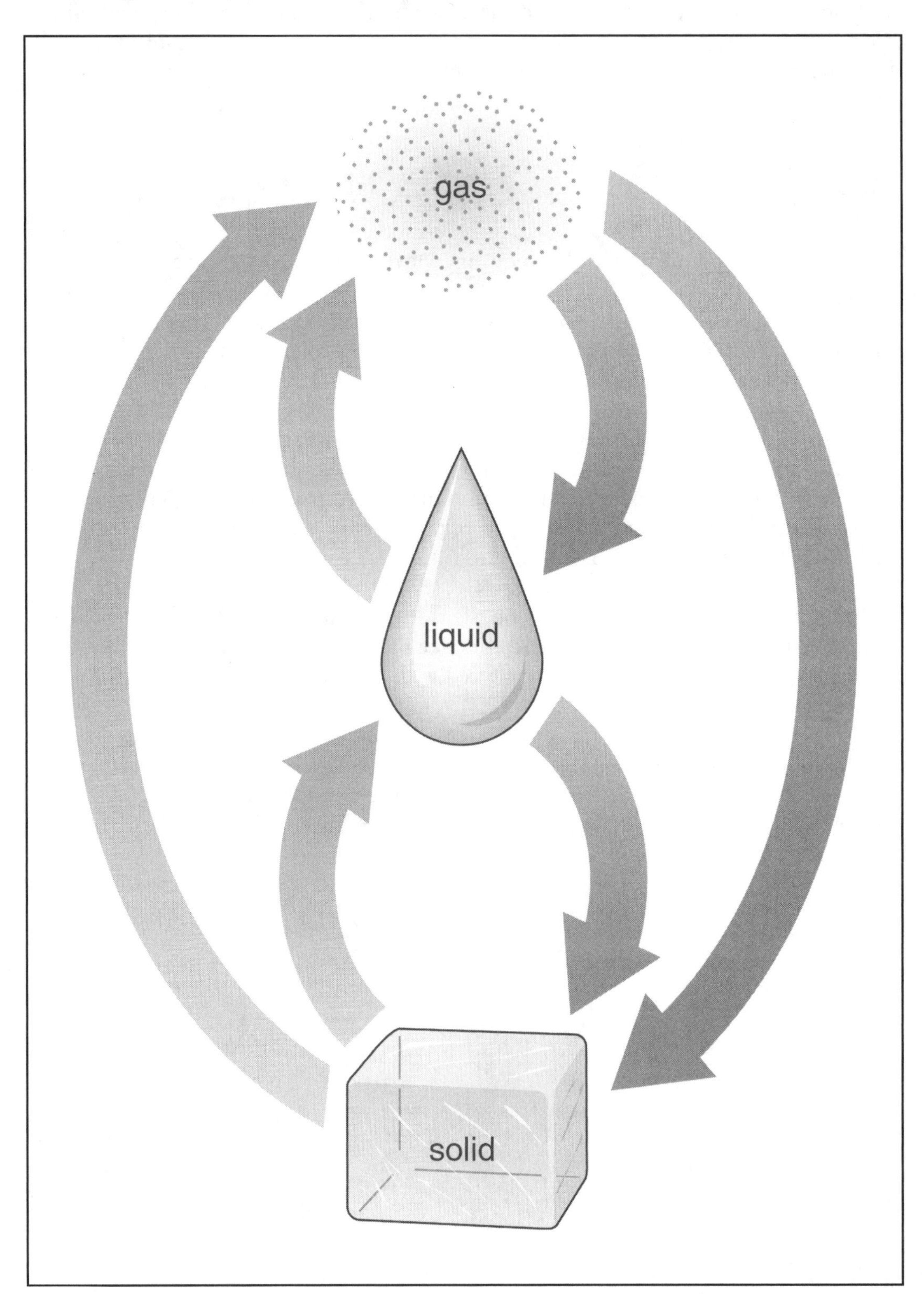

Blackline Master Ecology 7.2: The Water Cycle

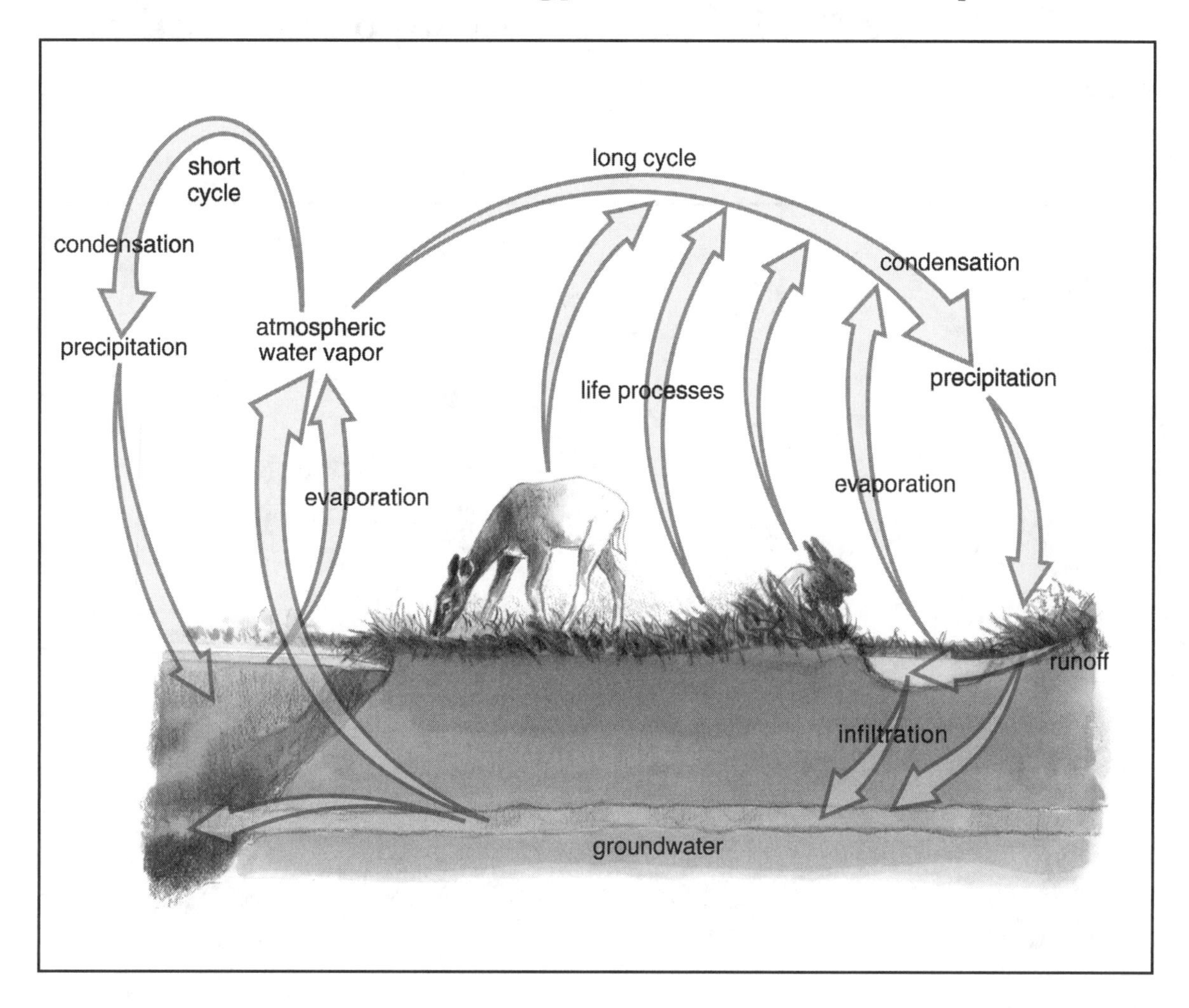

Blackline Master Ecology 7.3: The Water On Earth Goes Around and Around

Label the diagram of the water cycle. Use the following words (you may need to use some words more than once):

atmospheric water vapor
condensation
evaporation
groundwater
infiltration
life processes
long cycle
precipitation
runoff
short cycle

Blackline Master Ecology 7.4: Transpiration Experiment Table

Time (min)	Actual Reading in pipette (mL)	Adjusted Reading in pipette (mL)	Rate of Transpiration (mL/cm^2)
0			
3			
6			
9			
12			
15			
18			
21			
24			
27			
30			

Blackline Master Ecology 7.5: Area of Leaf Surfaces

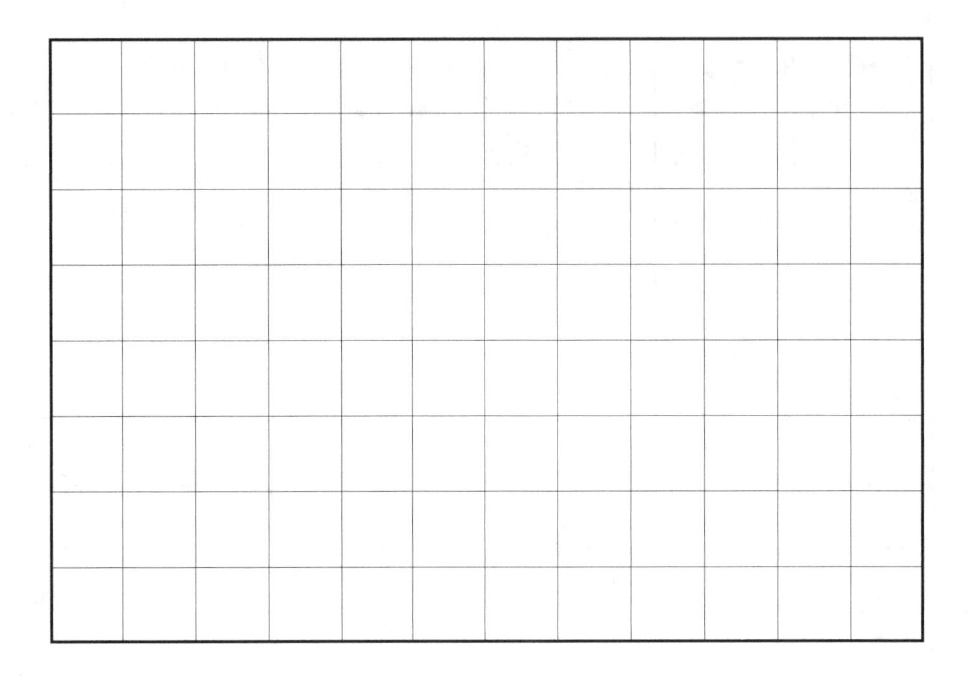

[metric graph 120 wide x 80 high / 10mm boxes]

Blackline Master Ecology 7.6: Rate of Transpiration

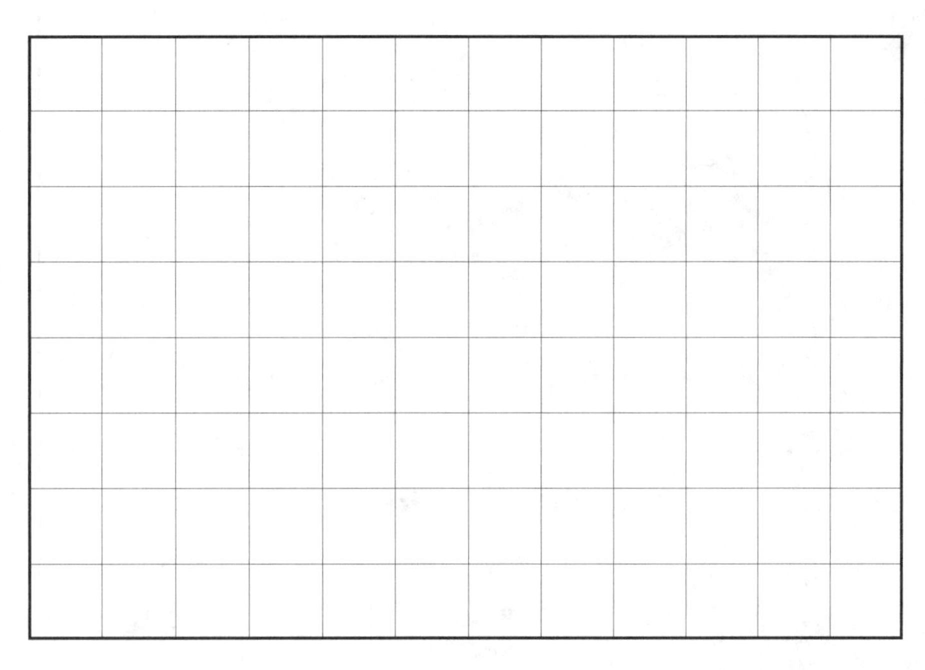

Blackline Master Ecology 7.7: Major Biomes of the Earth

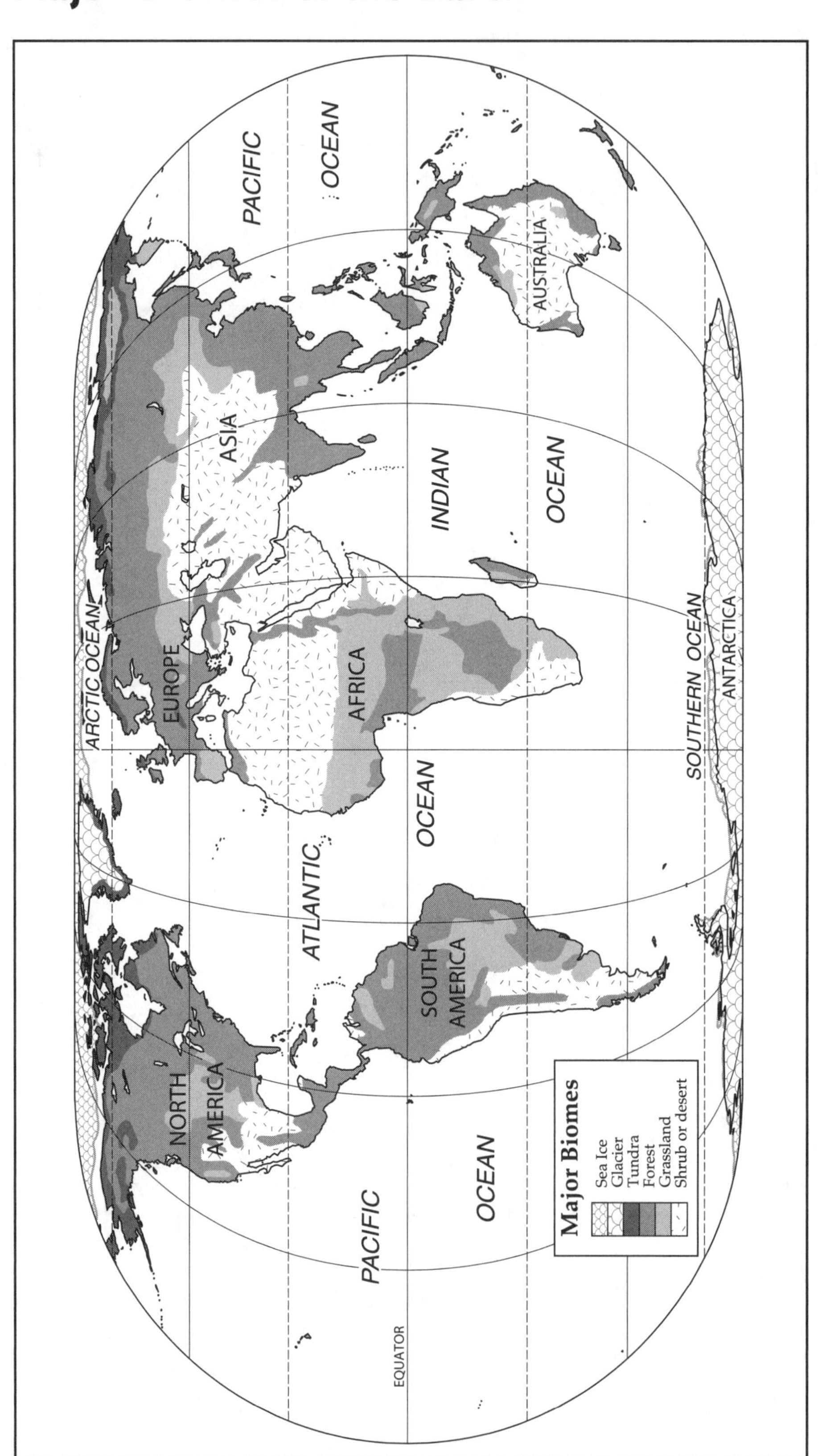

Blackline Master Ecology 7.8: Human Population Growth

Human Population Growth between A.D. 1 and 2000			
Date A.D.	Human Population (millions)	Date A.D.	Human Population (millions)
1	250	1930	2070
1000	280	1940	2300
1200	384	1950	2500
1500	427	1960	3000
1650	470	1970	3700
1750	694	1980	4450
1850	1100	1990	5300
1900	1600	2000	6080
1920	1800	2010	?

ACTIVITY 8–PHOTOSYNTHESIS, RESPIRATION, AND THE CARBON CYCLE

Background Information

Photosynthesis

Plants stay where they are and they cannot go anywhere to get food. They have to make their own "food" since they cannot get it from somewhere else. They use the energy from the Sun, the carbon from the carbon dioxide found in the atmosphere, and the water they absorbed through their roots to produce their own "food" in the form of starch or sugars. The process of doing this is called photosynthesis where carbon-carbon bonds are formed in the starch or sugars. Aside from the food that is produced, oxygen is also produced and is eventually released.

Respiration

With the sugars that they themselves made through photosynthesis, the plants now have a source of energy to do their own daily activities. They break the carbon-carbon bonds present in the sugars that they made and combine it with oxygen from the atmosphere to get energy, which is either used or lost, in the form of heat. This process releases carbon dioxide and water.

Organisms that cannot make their own food just do respiration and not photosynthesis. These living things get energy from the sugars that the plants made. They also get their energy by eating those that consume the plants. These organisms, just like the plants, break the carbon-carbon bonds to release the energy that they then can use and if not used is lost in the form of heat.

Carbon Cycle

The carbon that the plants took from the atmosphere in the form of carbon dioxide is changed into sugars when combined with water and light energy in the process of photosynthesis. With oxygen, the breakdown of carbon-carbon bonds present not only in the sugars that the plants made but also found in the bodies of those that accessed the sugars from the plants, results in the release of energy and with it carbon is returned to the atmosphere in the form of carbon dioxide. Together with the carbon dioxide, water is also released in the process of respiration.

Oxygen Cycle

Just like the carbon cycle, the oxygen cycle involves the processes of photosynthesis and respiration. In photosynthesis, carbon dioxide is used and oxygen is released. This oxygen is utilized in the process of respiration to release the energy stored in the sugars' carbon-carbon bonds. With the breakdown of the carbon-carbon bonds, energy is released and carbon dioxide is formed. The plants use the carbon dioxide to form the carbon-carbon bonds and with it oxygen is released. Oxygen shuttles between photosynthesis and respiration.

Goals and Assessments

Goal	Location in Activity	Assessment Opportunity
Learn how oxygen cycles through photosynthesis and respiration.	**For You To Do** **BioTalk**	**For You To Do** 1(a), 11(b), 11(c) and 14(b) **Biology to Go** Questions 1, 4(a) and 4(b)
Practice safe laboratory techniques for using chemicals in a laboratory situation.	**For You To Do**	**For You To Do** 2–15
Describe the cycling of carbon in an ecosystem.	**BioTalk**	**For You To Do** 1(a), 11(b), 11(c) and 14(b) **Biology to Go** Questions 2 and 3 **Inquiring Further**
Speculate how human activities can affect the carbon cycle.	**BioTalk**	**Biology to Go** Questions 3, 4(a) and 4(b) **Inquiring Further**

Activity Overview

This activity will show the role of sunlight in photosynthesis. With the use of the chemical iodine, the students are allowed to figure out whether or not starch is produced. This would lead them to answer whether or not photosynthesis occurred. Using petroleum jelly, a substance that the students are familiar with, they are encouraged to examine what part of the leaf allows for gas exchange – (carbon dioxide is taken in and oxygen is released) which is crucial for photosynthesis and respiration to happen.

Preparation and Materials Needed

Preparation

Two weeks before the experiment, you can start requesting the students to bring baby food jars together with their lids. They will be used for the isopropyl alcohol.

A week before the scheduled experiment, start looking for a plant to be used and be sure that you have them available no later than four days before the experiment. Since two plants of the same kind are used for each group, make sure that they are almost of the same size. The plants should have small leaves and at least eight of them. (Although the experiment calls for 5 leaves, allow 3 more leaves for possible mistakes.) It is advisable to get plants in a 4-inch pot. This will allow less space for exposure to light and darkness three days before the actual handling by the students.

If you do not have hot plates, you can use the microwave to provide the students with hot (60°C) tap water. Use Erlenmeyer flasks for easier manageability.

Materials/Equipment Needed (for the class)

- microwave
- Erlenmeyer flasks

(for each group)
Day 1

- 2 4-inch potted plants
- 1 pair of scissors
- 1 beaker
- 1 baby food jar with lid
- 2 forceps
- isopropyl alcohol
- hot (60°C) tap water
- petroleum jelly
- thermometer
- 4 sets of gloves
- 4 goggles
- 4 aprons
- soap
- paper towels

Day 2

- the baby food jar with the isopropyl alcohol from **Day 1**
- 1 baby food jar
- the same forceps used for the isopropyl alcohol from **Day 1**
- 1 forceps
- 2 Petri dishes
- Lugol's iodine solution
- white paper
- 4 sets of gloves
- 4 goggles
- 4 aprons
- soap
- paper towels

Day 4

- the same 4-inch potted plant originally kept in the dark and then was exposed to sunlight from **Day 1**
- Histoclear
- 8 swab applicators
- paper towels
- 1 baby food jar
- Lugol's iodine solution
- 4 Petri dishes
- 4 white papers
- 4 sets of gloves
- 4 goggles
- 4 aprons
- soap

Learning Strategies for Students with Limited English Proficiency

1. Point out new vocabulary in context. Practice using the words as much as possible.

flammable	toxic	expose	extracts
ingest	spill	splash	photosynthesis
mass	notch	label	giant redwood tree
carbon	exchange	detritus	carbon dioxide
decomposers	producers	consumers	carbon cycle
oxygen cycle	inorganic		

2. It is very important not only for the English Learners (ELs) but also for all the students to understand the safety precautions when dealing with chemicals in the laboratory. The words associated with safe handling of chemicals as well as those that should never be done should be explained very thoroughly.

3. For homework, have the students draw the procedures for the experiment. A visual representation always helps the students have a clearer understanding of the experiment. With their homework in hand, have the students in the groups that they are going to do the experiment compare drawings and adopt the best drawing that truly depicts the experimental procedures.

4. Have the students work in groups of four. Each student assumes a role from the four possible responsibilities – **reader**, **clarifier**, **questioner**, and **summarizer**. The **reader** reads what the book says and at the same time records what the group's answers are to the questions. The **clarifier** takes note of unfamiliar words that need to be made clear for the group to have a better understanding of the reading material. The **questioner** reminds the group of the questions that need to be answered like those found in **For You To Do**.

   This student also writes down any possible question that the group can think of. The **summarizer** draws the concepts that the group was able to learn in the activity.

NOTES

Activity 8 Photosynthesis, Respiration, and the Carbon Cycle

GOALS

In this activity you will:

- Learn how oxygen cycles through photosynthesis and respiration.
- Practice safe laboratory techniques for using chemicals in a laboratory situation.
- Describe the cycling of carbon in an ecosystem.
- Speculate how human activities can affect the carbon cycle.

What Do You Think?

Consider the mass of a seed from a giant redwood tree and the tree itself. It is hard to believe that a giant of a tree began as a small seed.

- **From where do the materials come to make up the mass of a mature tree?**

Write your answer to this question in your *Active Biology* log. Be prepared to discuss your ideas with your small group and other members of your class.

For You To Do

In this activity you will investigate what happens when the exchange of carbon dioxide between a leaf and the atmosphere is blocked.

1. Three days before this activity, one plant was placed in the dark. A second plant of the same species was placed in sunlight.

What Do You Think?

- The materials that make the mass of a mature tree come from the carbon in the carbon dioxide that it takes in through its leaves. This carbon dioxide, together with the water that it gets through its roots, plus the energy from the Sun allows the plant to manufacture its own "food" (starch or sugars) through photosynthesis. This is where the plant gets its energy to grow and maintain itself.

Student Conceptions

Students only remember plants for their ability to do photosynthesis. They seem to not include them to be able to do the process of respiration.

As to the concept of respiration, most students still associate this with breathing. Emphasize the point that it is through cellular respiration that we can access the energy from the food that we eat. Our cells go through cellular respiration using the oxygen that we took in through breathing.

Unlike energy, inorganic nutrients are recycled. When organisms die, decomposers extract the last of the available energy and some are lost in the form of heat. The inorganic nutrients – phosphorous, iron, nitrogen and others found in the dead bodies are then returned to the soil or water to be recycled again.

A Vote for Ecology

a) Predict what you will find when you test a leaf from each plant for the presence of starch.

Day 1

2. Remove one leaf from each plant. Use scissors to cut a small notch in the margin of the one placed in sunlight. Using forceps, drop the leaves into a beaker of hot (60°C) tap water.

Isopropyl alcohol is flammable and toxic. Do not expose the liquid or its vapors to heat or flame. Do not ingest; avoid skin/eye contact. In case of spills, flood the area with water and then call your teacher. Make sure you wear goggles, apron, and gloves

3. When the leaves are limp, use forceps to transfer the leaves to a screw-cap jar about half full of isopropyl alcohol. Label the jar with your team symbol and store it overnight as directed by your teacher.

4. Select four similar leaves on the plant that has been kept in the dark, but do not remove them from the plant. Using a fingertip, apply a thin film of petroleum jelly to the upper surface of one leaf. Check to be sure the entire surface is covered. (A layer of petroleum jelly, although transparent, is a highly effective barrier across which many gases cannot pass.) Cut one notch in the leaf's margin.

5. Apply a thin film to the lower surface of a second leaf and cut two notches in its margin.

6. Apply a thin film to both upper and lower surfaces of a third leaf and cut three notches in its margin.

7. Do not apply petroleum jelly to the fourth leaf, but cut four notches in its margin; Place the plant in sunlight.

a) What is the purpose of the leaf marked with four notches?

8. Wash your hands thoroughly before leaving the laboratory.

Day 2

9. Obtain your jar of leaf-containing alcohol from Day 1. Using forceps, carefully remove the leaves from the alcohol and place them in a beaker of room-temperature water. (The alcohol extracts chlorophyll from the leaves but also removes most of the water, making them brittle.) Recap the jar of alcohol and return it to your teacher.

10. When the leaves have softened, place them in a screw-cap jar about half full of Lugol's iodine solution.

Lugol's iodine solution is used to test for the presence of small amounts of starch. Starch gives a blue-black color.

11. After several minutes, use forceps to remove both leaves, rinse them in a beaker of water, and spread them out in open Petri dishes of water placed on a sheet of white paper.

a) Record the color of each leaf. Recap the jar of Lugol's iodine solution and return it to your teacher.

b) What was the purpose of the iodine test on Day 2?

c) If you use these tests as an indication of photosynthetic activity, what are you assuming?

Lugol's iodine solution is poisonous if ingested, irritating to skin and eyes, and can stain clothing. Should a spill or splash occur, call your teacher immediately; flush the area with water for 15 minutes; rinse mouth with water.

For You To Do
Teaching Suggestions and Sample Answers

1. a) Student answers will vary. The plant kept in the dark for three days will have less starch content than the plant exposed to sunlight.

7. a) The leaf marked with four notches is the control for the experiment. It will be used as a standard to compare the results using the other leaves.

11. b) The iodine test on **Day 2** was to determine the role of sunlight in the production of starch.

12. Wash your hands thoroughly before leaving the laboratory.

Day 4

13. Remove from the plant the four notched leaves prepared on Day 1 and place them on paper towels. To remove the petroleum jelly, dip a swab applicator in the Histoclear^{TM} and gently rub it over the surface of the film once or twice. Then gently use a paper towel to remove any residue of petroleum jelly. Discard the swab applicator and the paper towel in the waste bag.

14. Repeat **Steps 10** and **11.**
    a) Compare the color reactions of the four leaves and record your observations.
    b) In which of the leaves coated with petroleum jelly did photosynthetic activity appear to have been greatest? Least?

15. Wash your hands thoroughly before leaving the laboratory.

Histoclear is a combustible liquid. Do not expose to heat or flame. Do not ingest; avoid skin/eye contact. Should a spill or splash occur, call your teacher immediately; wash skin area with soap and water.

BioTalk

The Carbon Cycle

You take in carbon in all the foods you eat. You return carbon dioxide to the air every time you exhale. A plant also returns carbon dioxide to the air when it uses its own sugars as a source of energy. When another plant takes in the carbon dioxide during photosynthesis, the cycle of carbon through the community is complete. In this activity you observed what happens when this exchange of carbon dioxide does not take place. However, when the exchange does take place, the plant can use the carbon from the carbon dioxide to live and grow.

Carbon dioxide is also returned to the air by decomposers. When producers or consumers die, decomposers begin their work. As its source of energy, a decomposer uses the energy locked in the bodies of dead organisms. It uses the carbon from the bodies to build its own body. Carbon that is not used is returned to the air as carbon dioxide. Eventually, almost all the carbon that is taken in by plants during photosynthesis is returned to the air by the activity of decomposers.

14. b) The leaf coated with petroleum jelly on its upper surface has the greatest amount of photosynthetic activity as shown by its more pronounced blue-black coloration. The leaf with its lower surface coated and the leaf with both of its upper and lower surface coated had the least photosynthetic activity as shown by the lack of blue-black coloration with the iodine test.

Teaching Tip

Using **Blackline Master Ecology 8.4: The Structure of a Leaf,** ask the students where gas exchange is possible. Also ask them to explain their response. Connect their experimental results with the structure of the leaf. Have them figure out what part of the leaf should not be coated with the petroleum jelly to allow carbon dioxide to get to the leaf.

A Vote for Ecology

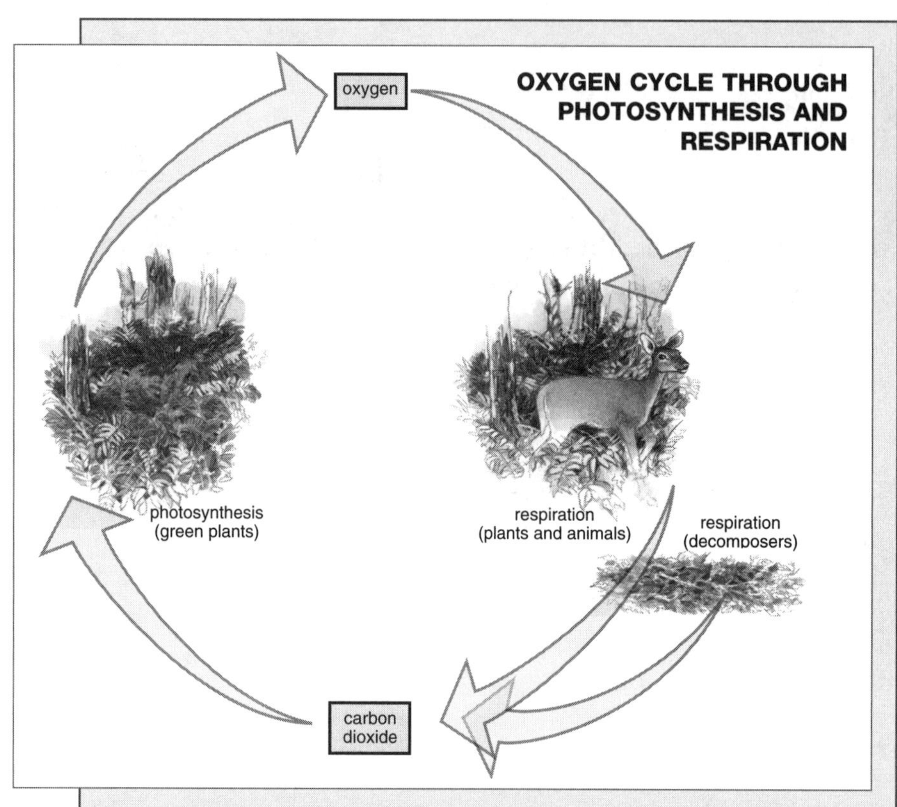

Hundreds of millions of years ago, many energy-rich plant bodies were buried before decomposers could get to them. When that happened, the bodies slowly changed during long periods of time. They became a source of fuels like coal, oil, and natural gas. Today, when these fuels are burned, energy is released. The carbon in the fuels is returned to the air as carbon dioxide. You can see that even the energy obtained from fuels is a result of photosynthesis. The process in which carbon is passed from one organism to another, then to the abiotic community, and finally back to the plants is called the **carbon cycle**.

The Cycling of Matter

The energy from the Sun flows through the ecosystem in the form of carbon-carbon bonds in organic matter. When respiration occurs, the carbon-carbon bonds are broken and energy is released. The carbon is

BioTalk

Teaching Tips

Draw W ↑ E on the board. (N at top, S at bottom)

Then discuss the oxygen cycle by pointing out what process is found on the north; on the south; on the west; and on the east. Have the students note the direction of the arrows. Is it clockwise or counterclockwise?
(Clockwise)

The oxygen cycle is available as **Blackline Master 8.1.**

Assessment Opportunity

The following questions can be used to assess the students' understanding of the oxygen cycle:

1. Give one similarity between the materials that are found on the north and south axes of the oxygen cycle.

   (They are both products of a process; they are both gases)
2. Give one difference between the materials that are found on the north and south axes of the oxygen cycle.

   (Oxygen is given off in photosynthesis while carbon dioxide is used in photosynthesis; carbon dioxide is given off in respiration while oxygen is used in respiration.)
3. Give one similarity between the processes that are found on the west and east axes of the oxygen cycle.

   (Photosynthesis and respiration are life processes)
4. Give one difference between the processes that are found on the west and east axes of the oxygen cycle.

   (Photosynthesis is used only by the plants while respiration is a process used by both plants and animals, including decomposers)
5. Explain why this is a cycle.

   (Photosynthesis releases oxygen that is used in respiration. Respiration then releases carbon dioxide that is used in photosynthesis and in the process produces oxygen. Oxygen again is taken in for respiration to occur, and the cycle keeps going.)

combined with oxygen to form carbon dioxide. The energy that is released is either used by the organism (to move, digest food, excrete wastes, etc.) or the energy may be lost as heat. In photosynthesis energy is used to combine the carbon molecules from the carbon dioxide, and oxygen is released. This is illustrated in the diagram. All the energy comes from the Sun. The ultimate fate of all energy in ecosystems is to be lost as heat. Energy does not recycle!

However, inorganic nutrients do recycle. They are inorganic because they do not contain carbon-carbon bonds. These inorganic nutrients include the phosphorous in your teeth, bones, and cell membranes. Also, nitrogen is found in your amino acids (the building blocks of protein). Iron is in your blood. These are just a few of the inorganic nutrients found in your body. Autotrophs obtain these inorganic nutrients from the inorganic nutrient pool. These nutrients can usually be found in the soil or water surrounding the plants or algae. These inorganic nutrients are then passed from organism to organism as one organism is consumed by another. Ultimately, all organisms die. They become detritus, food for the decomposers. At this stage, the last of the energy is extracted (and lost as heat). The inorganic nutrients are returned to the soil or water to be taken up again. The inorganic nutrients are recycled; the energy is not.

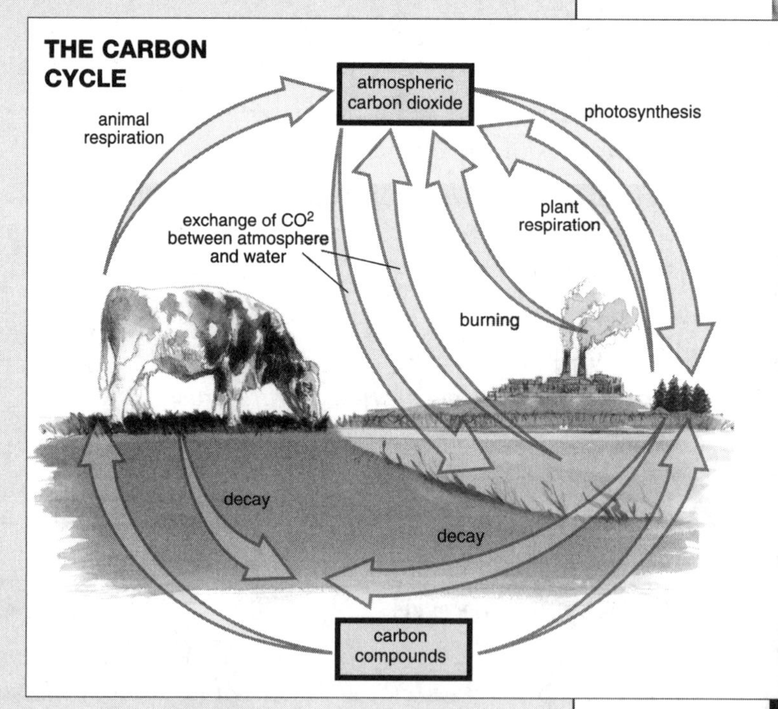

Bio Words

carbon cycle: the process in which carbon is passed from one organism to another, then to the abiotic community, and finally back to the plants

Teaching Tips

With the overwhelming contributions to the atmospheric carbon dioxide by animal respiration, CO_2 from the water, burning, and plant respiration, you would expect the atmospheric CO_2 to be very high. This is not the case since the CO_2 is used by the photosynthesizers and is absorbed by the oceans forming dissolved CO_2 in the process. Atmospheric CO_2 makes up only about 0.03% of the atmosphere. The carbon available in the ocean is fifty times that found in the atmosphere.

The carbon cycle is available as **Blackline Masters 8.2 and 8.3.**

Reflecting on the Activity and the Challenge

In this activity you learned that carbon is the key element in all organic matter. You investigated the process of photosynthesis and then related this process to respiration in the carbon-oxygen cycle. The cycling of matter like carbon is essential to the survival of any ecosystem. You will need to explain this cycle in your booklet.

Biology to Go

1. Explain why photosynthesis and cellular respiration are considered to be paired processes.
2. What is the importance of decomposers in the carbon cycle?
3. What effect does the burning of fossil fuels have on the carbon cycle?
4. Scientists have expressed concerns about the burning of the rainforests to clear the land for the planting of crops.
   a) Explain how the burning of the forests could change oxygen levels.
   b) What impact would the change in oxygen levels have on living things?

Inquiring Further

The greenhouse effect

The term greenhouse effect was coined in the 1930s to describe the heat-blocking action of atmospheric gases. Research and report the connection between the greenhouse effect and the carbon cycle.

Reflecting on the Activity and the Challenge

Teaching Tips

This is the perfect time to ask the students, in their groups, to discuss what they learned in this activity. After a 3-minute discussion within their groups, have a class discussion where all the groups are expected to give one concept that they understood from this specific activity.

After the class discussion, remind the students that these concepts are to be included in their **Chapter Challenge** of making a booklet addressing or focusing on one environmental/ecological issue.

A Vote for Ecology

Reflecting on the Activity and the Challenge

In this activity you learned that carbon is the key element in all organic matter. You investigated the process of photosynthesis and then related this process to respiration in the carbon-oxygen cycle. The cycling of matter like carbon is essential to the survival of any ecosystem. You will need to explain this cycle in your booklet.

Biology to Go

1. Explain why photosynthesis and cellular respiration are considered to be paired processes.
2. What is the importance of decomposers in the carbon cycle?
3. What effect does the burning of fossil fuels have on the carbon cycle?
4. Scientists have expressed concerns about the burning of the rainforests to clear the land for the planting of crops.
   a) Explain how the burning of the forests could change oxygen levels.
   b) What impact would the change in oxygen levels have on living things?

Inquiring Further

The greenhouse effect

The term greenhouse effect was coined in the 1930s to describe the heat-blocking action of atmospheric gases. Research and report the connection between the greenhouse effect and the carbon cycle.

Biology to Go

1. Photosynthesis releases oxygen that is needed in cellular respiration. Cellular respiration, on the other hand, releases carbon dioxide, which is needed for photosynthesis to occur.

2. Decomposers use the energy from the carbon-carbon bonds found in the dead bodies of plants and animals. It uses the carbon from the bodies to maintain its very own body. The carbon that is not used is returned to the atmosphere as carbon dioxide.

3. The burning of the fossil fuels releases energy and the carbon in the fuels is returned to the atmosphere in the form of carbon dioxide.

4. a) The burning of the forests will result in less photosynthetic activity. With decreased photosynthesis, oxygen production also decreases and oxygen level is lower.

   b) With the decrease in oxygen levels, all living things will be negatively affected since cellular respiration needs oxygen to access the energy in the carbon-carbon bonds.

Inquiring Further

The greenhouse effect
Increased burning of fossil fuel adds to the atmospheric CO_2. With high levels of atmospheric CO_2, greenhouse effect occurs where infrared radiation in the atmosphere is trapped resulting in higher temperature on Earth.

Assessment Opportunity

Assign **Inquiring Further** as homework. Specify that students are to access the Internet for this endeavor. Require them to cite at least three web sites as their sources.

Blackline Master Ecology 8.1: The Oxygen Cycle

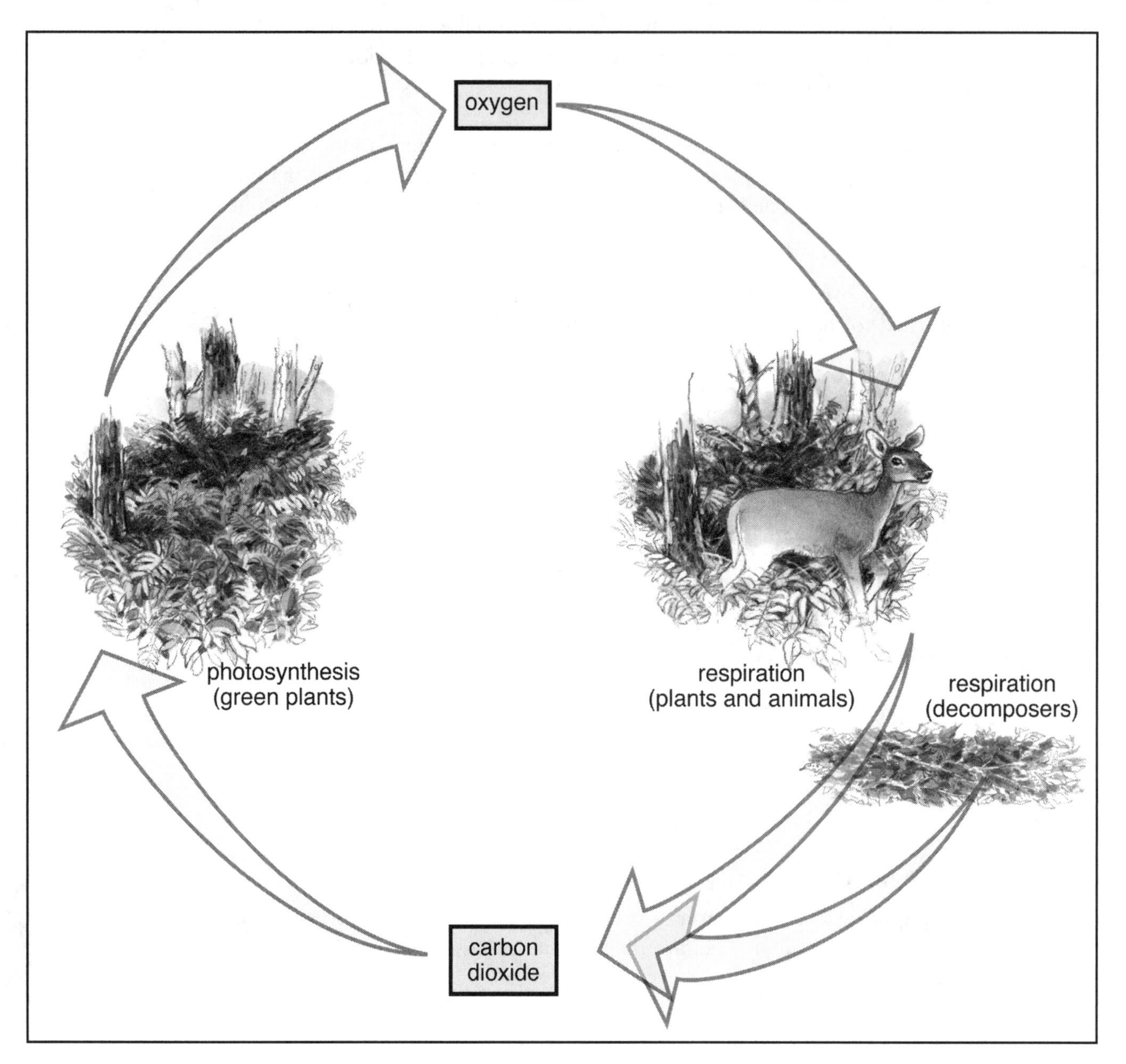

Blackline Master Ecology 8.2: The Carbon Cycle

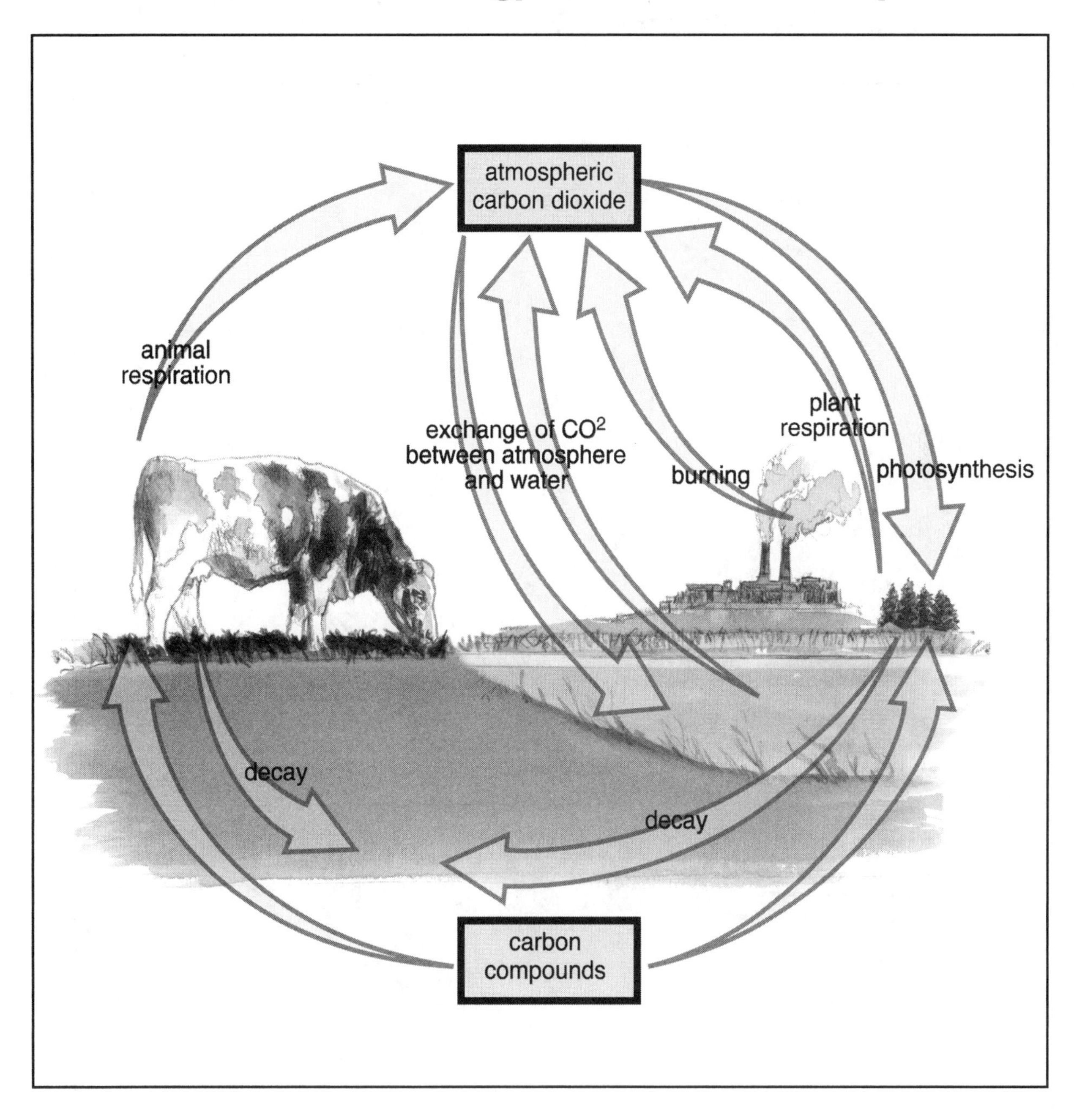

Blackline Master Ecology 8.3: Carbon Is Recycled on Earth

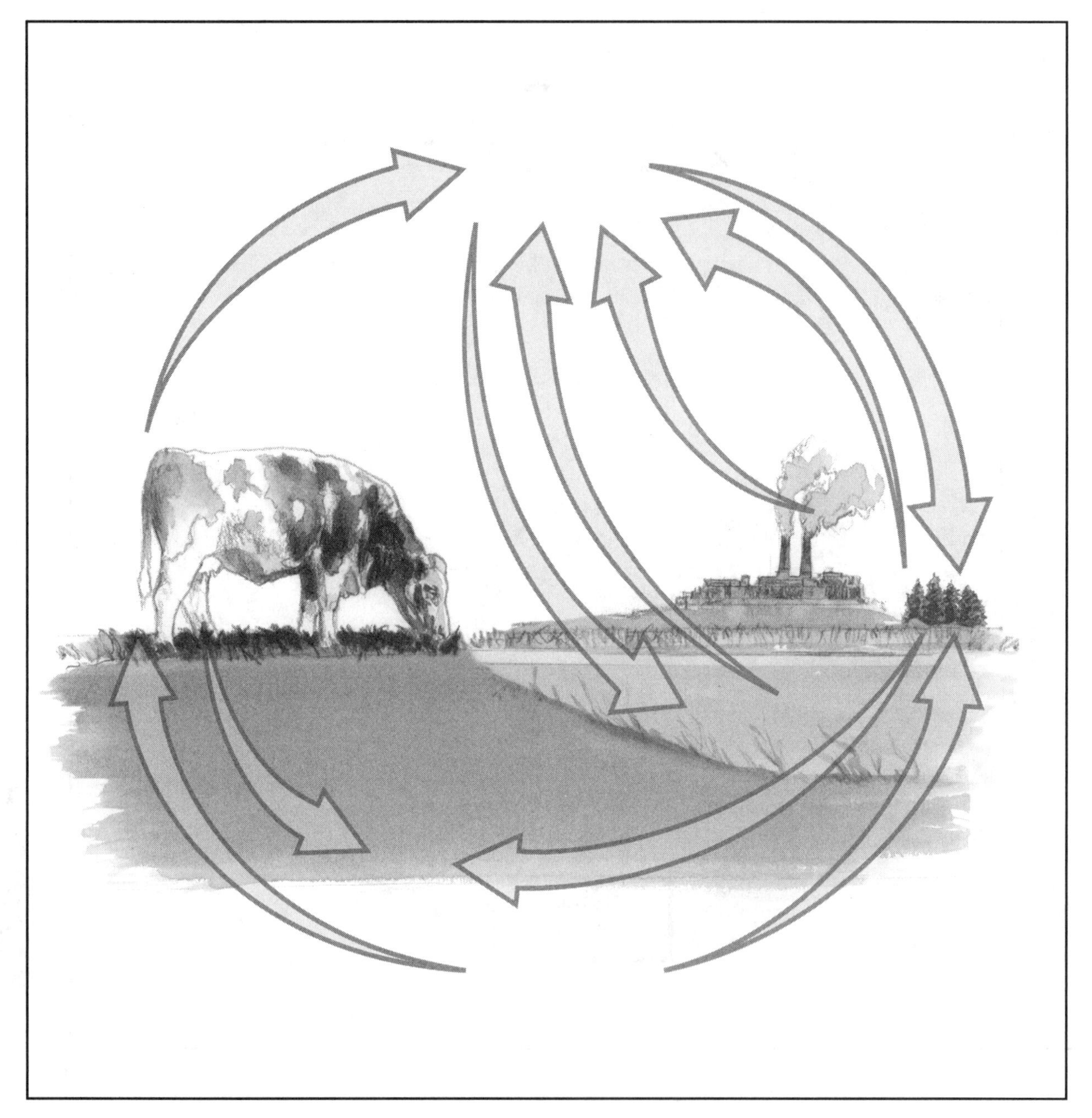

Label the diagram of the carbon cycle. Use the following words or phrases:

animal respiration
atmospheric carbon dioxide
burning
carbon compounds
decay
exchange of carbon dioxide between atmosphere and water
photosynthesis
plant respiration

Blackline Master Ecology 8.4: The Structure of a Leaf

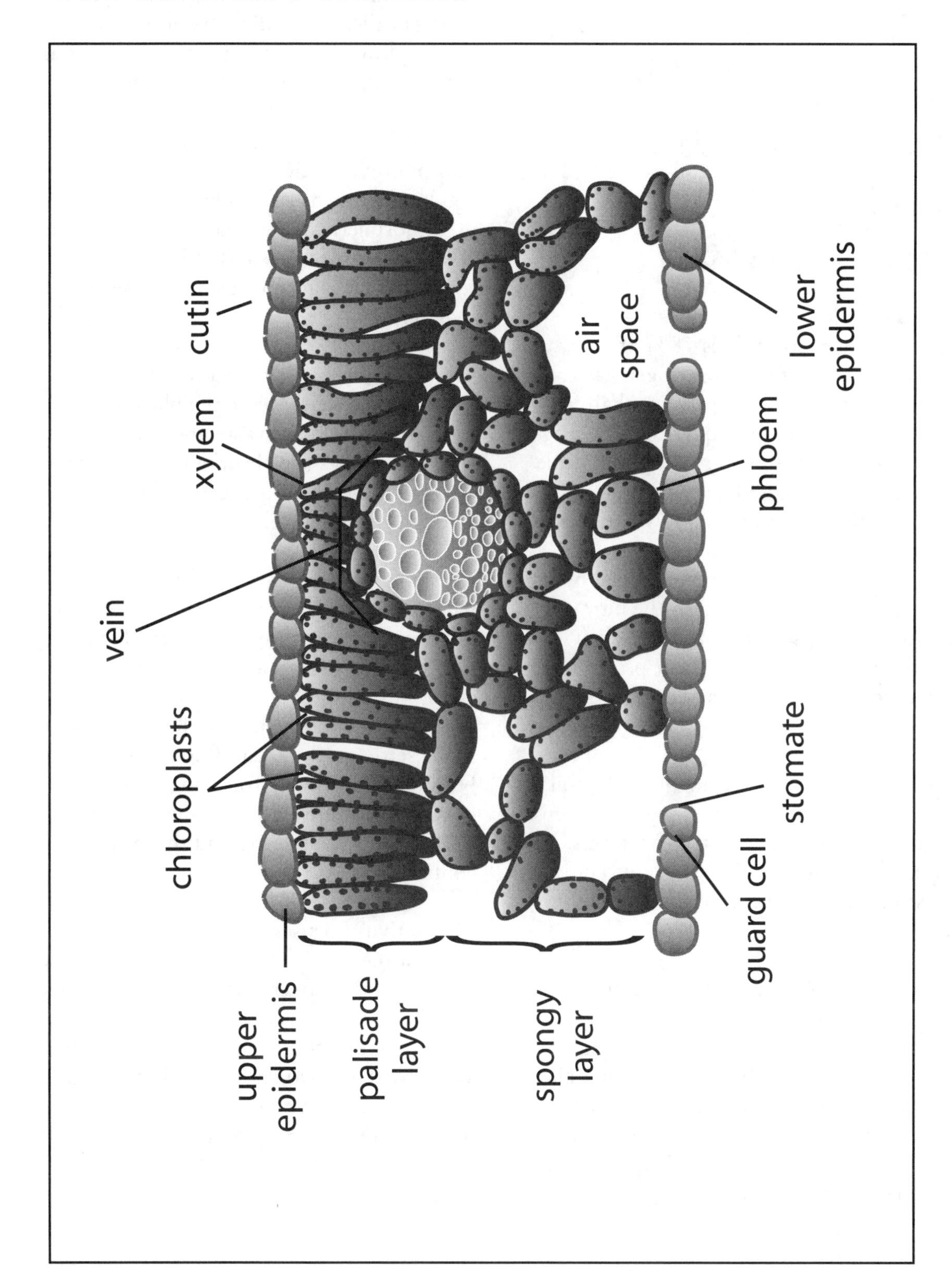

ACTIVITY 9– THE NITROGEN AND PHOSPHOROUS CYCLES

Background Information

Nitrogen Cycle

Even if nitrogen constitutes 78% of the atmosphere in the form of nitrogen gas (N_2), it is not accessible to living things that need nitrogen for protein or deoxyribonucleic acid (DNA). N_2 or atmospheric nitrogen is made available only through lightning and soil-bacteria action. Lightning allows N_2 to react with oxygen-forming nitrates (NO_3-) that enter the soil where they are absorbed by the plants' roots. Plants transform the nitrates into amino acids to form protein. When animals eat the plants, the plant proteins are broken down into amino acids and then formed into animal proteins.

Another way to access N_2 involves the soil bacteria (**nitrogen-fixing bacteria**) found in root nodules of legumes like peas, beans, and clover. These bacteria change N_2 into nitrates. Excess nitrates produced go into the soil for other plants to use.

Oxygen-using decomposers or bacteria (**aerobic bacteria**) change the nitrogen-containing chemicals in dead bodies and waste into ammonia (NH_3). This ammonia is transformed into nitrites (NO_2) by other bacteria and still others convert the nitrites back into nitrates, which are then absorbed by the plants' roots.

Bacteria that do not need oxygen (**anaerobic bacteria**) do the opposite. They change the nitrates into nitrites and back into N_2. This process called **denitrification** ensures the return of nitrogen into the atmosphere in the form of nitrogen gas, making the cycle complete.

Phosphorous Cycle

Unlike the other cycles, phosphorous has no atmospheric counterpart. It is needed in DNA, cell molecules, and in bones and teeth. It cycles in two ways–short term and long term.

The short but faster term involves decomposers that release the phosphates as they break down waste and dead tissue. The phosphates are absorbed by the plants through their roots and are then cycled through the food webs.

The long term involves millions of years where weathering of rocks and leaching result in phosphates reaching the ocean. Here they may be absorbed by aquatic organisms and is cycled through the food webs. With the death of these aquatic organisms, their bones and shells settle on the ocean floor and eventually become rocks. These rocks may become exposed by uplifting where they are then subjected to weathering and the cycle continues.

This activity focuses on the observational skills of the students as they experimentally investigate the effect on the growth of algae by a lawn fertilizer and a detergent. The students are encouraged to figure out how and why these two chemicals that are known to be "good" can actually be "bad" where other organisms are concerned.

Goals and Assessments

Goal	Location in Activity	Assessment Opportunity
Investigate the chemicals that promote and inhibit the growth of plant material.	For You To Do	**For You To Do** 1(a), 4(a), 7(a) and 11(b) **Biology to Go** Question 1
Explain the importance of nitrogen and phosphorous to organisms.	BioTalk	**Biology to Go** Questions 1, 5 and 6(b)
Describe how nitrogen cycles in an ecosystem.	BioTalk	**BioTalk** **Biology to Go** Questions 2, 3, 4 and 6(b)
Describe how phosphorous cycles in an ecosystem.	BioTalk	**BioTalk**
Provide examples of how human activities can affect the nitrogen cycle.	BioTalk	**BioTalk** **Biology to Go** Questions 6(a), 6(b), and 7 **Inquiring Further** Question 2

Activity Overview

This activity provides an opportunity for the students to learn about the nitrogen and the phosphorous cycles. The students conduct an experiment measuring the effect of lawn fertilizer and detergent on the growth of algae. In this activity, using common household chemicals, it is hoped that the students will become more cognizant of human actions as they impact not only the nitrogen and phosphorous cycles but organisms as well.

Preparation and Materials Needed

Preparation

Two weeks before the experiment, you can start reminding the students to start bringing detergent to school. Since there are usually 8 groups in a class, 4 kinds of detergent can be used where two groups will use the same detergent. The groups can then compare results and see which detergent allows for the highest algal growth.

As to the fertilizer, get one that has the same nitrogen and phosphorous content. This makes for fewer variables to consider.

Wide-mouth jars are easier to clean and are preferred for more efficient exchange of oxygen with the air.

Materials/Equipment Needed (for each group)

- 3 1-L jars
- 2.25 L distilled water (750 mL per jar
- 30-mL pond water with algae (10 mL per jar)
- 45 gm lawn fertilizer (15 gm in 1 jar)
- 45 gm detergent (15 gm in 1 jar)
- plastic wrap
- glass marker
- 3 filter papers
- 3 funnels
- scale

(for each student)

- gloves
- goggles
- apron
- soap
- paper towels

Learning Strategies for Students with Limited English Proficiency

1. Point out new vocabulary in context. Practice using the words as much as possible.

peril	promote	detergent	phosphorous cycle
filtrate	nitrogen cycle	converted	rotating crops
nodules	denitrification	nitrates	algal bloom
nitrites	weathering	sediment	nitrogen fixation
algae	inorganic	detrivores	building block
detritus	flourish	uplifting	mixed plantings of crops
organic	fertilizer		

2. For homework:
   - Have the students draw the steps in the experiment. This pictorial representation will help them in knowing what to do during the experiment itself.

Activity 9 The Nitrogen and Phosphorous Cycles

GOALS

In this activity you will:

- Investigate the chemicals that promote and inhibit the growth of plant material.
- Explain the importance of nitrogen and phosphorous to organisms.
- Describe how nitrogen cycles in an ecosystem.
- Describe how phosphorous cycles in an ecosystem.
- Provide examples of how human activities can affect the nitrogen cycle.

What Do You Think?

Nitrogen is essential to all forms of life. Yet, recent studies have shown that excess nitrogen has been introduced into our ecosystems. It has had negative effects on the natural nitrogen cycle.

- **What are the sources of the excess nitrogen?**
- **What are some of the negative effects of too much nitrogen?**

Write your answer to these questions in your *Active Biology* log. Be prepared to discuss your ideas with your small group and other members of your class.

For You To Do

An excessive growth of algae (algal blooms) can make a lake very unappealing. More importantly, it places other organisms in the ecosystem in peril through lack of oxygen. In this activity you will investigate some of the chemicals that promote the growth of algae.

What Do You Think?

- Excess nitrogen comes from fertilizers and car exhaust.

- Too much nitrogen results in overproduction of algae known as algal bloom. Depletion of the oxygen supply occurs and this results in the death of other organisms.

Student Conceptions

Students think that since detergents kill germs then necessarily they would also kill algae. Another misconception would be that since fertilizers are good for plants then it must also be good for algae.

This may be the first time that students will get introduced to the concept of beneficial bacteria. They have been exposed to the idea that bacteria are harmful.

A Vote for Ecology

Handle all of the liquids and chemicals very carefully. They should all be considered contaminated and toxic. Keep hands away from eyes and mouth during the activity. Wash your hands well after the activity. Clean up any spills immediately.

1. Obtain three 1-L jars. Make sure the jars are rinsed thoroughly, so that there are no leftover traces of any chemicals, including soap. Fill each jar about three-fourths full with distilled water.
   a) Why is it important that the jars be cleaned before beginning this activity?
2. To each jar add a 10-mL sample of pond water. Stir the pond water thoroughly before taking the sample. The pond water will contain algae.
3. Label the jars A through C.
4. To each jar add the following:
   - To Jar A, add 15 gm of detergent.
   - To Jar B, add 15 gm of lawn fertilizer.
   - Do not add anything to Jar C.

   a) Many detergents contain phosphates. Fertilizers contain nitrogen and phosphates. Write as a question what you are investigating in this activity.
   b) What is the purpose of Jar C?
5. Cover each jar with plastic wrap so that dirt will not settle into the jar, but allow for some air to enter the jar.
6. Use a glass marker to mark the water level in each jar.
7. Set all the jars in a well-lighted place, but not in direct sunlight.
   a) Predict in which jar the algal growth will be the greatest? The least? Give reasons for your predictions.
8. Observe the jars each day for about two weeks. As water evaporates from the jars, add distilled water to bring the water back up to its original level. At the end of two weeks, you will pass the water in each jar through a separate filter.
   a) Record your observations every two or three days.
9. Find the mass of each of three pieces of filter paper.
   a) Record the mass of each in a table.
10. Fold the filter paper as shown and insert it into a funnel. Place the funnel in the mouth of another jar to collect the filtrate (the liquid that passes through the filter).

    Filter the liquid in each of the three jars.
11. Allow the filter papers and the algae residue to dry thoroughly.

    Find the mass of each piece of filter paper and algae. Calculate the mass of the algae.

    a) Record your findings in a table.
    b) Did your findings support your predictions? Explain any differences you found.

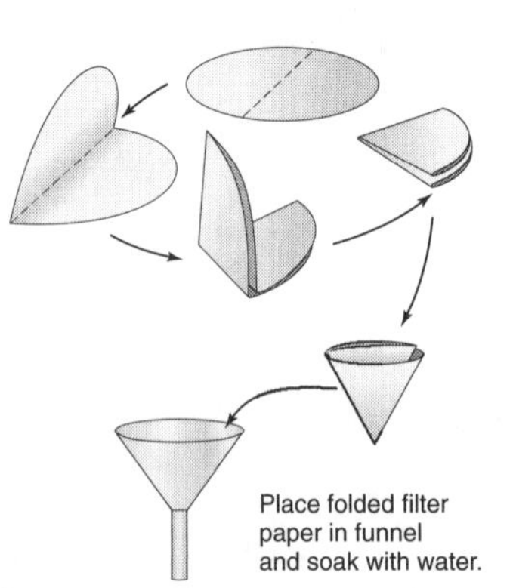

Be very careful with the liquid and the algae residue. You should assume that disease organisms have grown in the water during the activity. Be very careful to avoid ingesting any of the water or residue. Dispose of all materials as directed by your teacher when finished.

For You To Do
Teaching Suggestions and Sample Answers

1. a) Without thorough cleaning, contaminants may be inadvertently introduced into the experiment.

4. a) How do phosphates affect algal growth?
How do phosphates and nitrates affect algal growth?

b) Jar C is the control.

7. a) Algal growth will be greatest in Jar B with fertilizer since both phosphates and nitrates that are needed by the algae are present in huge amounts. Algal growth will be least in Jar C since phosphates and nitrogen are not present in huge amounts.

11. b) Student answers may vary. The findings support what has been predicted since both phosphates and nitrogen, which are necessary for algal growth and development, are present in the fertilizer.

Teaching Tip

Using **Blackline Master Ecology 9.3: Effect of Detergent and Lawn Fertilizer on Algal Growth** discuss with the students what characteristics they should look for in their jars to describe the developments that are occurring. Emphasize the importance of keeping good records. To prevent loss of experimental results, require the groups not to have just one student keep records but that every single member should do so.

Bio Talk

THE NITROGEN CYCLE

Nitrogen Fixation

Nitrogen is a basic building block of plant and animal proteins. It is a nutrient essential to all forms of life. Nitrogen is also required to make deoxyribonucleic acid or DNA. DNA is the hereditary material found in all living things. The movement of nitrogen through ecosystems, the soil, and the atmosphere is called the **nitrogen cycle**. Like carbon, nitrogen moves in a cycle through ecosystems. It passes through food chains and from living things to their environment and back again. Life depends on the cycling of nitrogen.

The largest single source of nitrogen is the atmosphere. It is made up of 78 percent of this colorless, odorless, nontoxic gas. With this much nitrogen available, you would think organisms would have no difficulty getting nitrogen. Unfortunately, this is not the case. Nitrogen gas is a very stable molecule. It reacts only under limited conditions. In order to be useful to organisms, nitrogen must be supplied in another form, the nitrate ion (NO_3-).

Three processes are responsible for most of the nitrogen fixation in the biosphere: atmospheric fixation by lightning, biological fixation by certain microbes, and industrial fixation. The enormous energy of lightning breaks nitrogen molecules apart. Only about five percent of the nitrates produced by nitrogen fixation are produced by lightning.

The nitrogen cycle is very complex. A simplified description is shown in the diagram on the next page. There are two ways in which atmospheric nitrogen can be converted into nitrates. The first method is lightning, and the second is bacteria in the soil. The process of converting nitrogen into nitrates is called **nitrogen fixation**.

A small amount of nitrogen is fixed into nitrates by lightning. The energy from lightning causes nitrogen gas to react with oxygen in the air, producing nitrates. The nitrates dissolve in rain, falling to Earth and forming surface water.

Bio Words

nitrogen cycle: the movement of nitrogen through ecosystems, the soil, and the atmosphere

nitrogen fixation: the process by which certain organisms produce nitrogen compounds from the gaseous nitrogen in the atmosphere

BioTalk

Teaching Tip

Ask the students what purpose lightning has. Students may not know the role that lightning has in the nitrogen cycle. Tell them that the lightning's energy destabilizes the atmospheric nitrogen, making it react with oxygen to form NO_3- that can then be used by the plants.

The nitrates enter the soil and then move into plants through their roots. Plant cells can use nitrates to make DNA, and they can convert nitrates into amino acids, which they then string together to make proteins. When a plant is consumed by an animal, the animal breaks down the plant proteins into amino acids. The animal can then use the amino acids to make the proteins it needs.

THE NITROGEN CYCLE

nitrogen gas (atmosphere)
nitrogen fixation (lightning)
nitrogen in protein and other organic compounds
feeding
artificial fertilizers
nitrogen fixation (bacteria)
denitrifying bacteria
waste and death
taken up by plants
decomposers
nitrogen in inorganic forms

Some bacteria are capable of fixing nitrogen. These bacteria provide the vast majority of nitrates found in ecosystems. They are found mostly in soil, and in small lumps called nodules on the roots of legumes such as clover, soybeans, peas, and alfalfa. The bacteria provide the plant with a built-in supply of usable nitrogen, while the plant

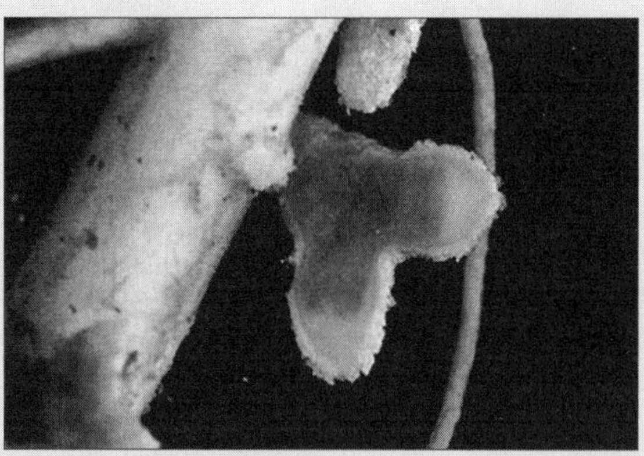

The most familiar examples of biotic nitrogen fixing are the root nodules of legumes, plants like peas, beans, and clover.

Teaching Tips

Before discussing the nitrogen cycle, assign **The Nitrogen Cycle** (pp. 599-602) in the **BioTalk** reading for homework.

Ask the students to make **Cornell Notes** on those pages.

Encourage them to have their homework completed by telling them that they can use their notes in a quiz the next day.

Assessment Opportunity

Using **Blackline Master 9.1: The Nitrogen Cycle**, have the students work as partners by making two students who are seated close to each other move their desks so that they are facing each other. Give the partners two pairs of scissors, glue, and a white paper. Have them construct **The Nitrogen Cycle** with the aid of their **Cornell Notes**.

supplies the nitrogen-fixing bacteria with the sugar they need to make the nitrates. This plant-bacteria combination usually makes much more nitrate than the plant or bacteria need. The excess moves into the soil, providing a source of nitrogen for other plants. The traditional agricultural practices of rotating crops and mixed plantings of crops, one of which is always a legume, capitalizes on bacterial nitrogen fixation.

All organisms produce wastes and eventually die. When they do, decomposers break down the nitrogen-containing chemicals in the waste or body into simpler chemicals such as ammonia (NH_3). Other bacteria convert ammonia into nitrites, and still others convert the nitrites back to nitrates. These bacteria all require oxygen to function. The nitrates then continue the cycle when they are absorbed by plant roots and converted into cell proteins and DNA.

Farmers and gardeners who use manure and other decaying matter take advantage of the nitrogen cycle. Soil bacteria convert the decomposing protein in the manure into nitrates. Eventually, the nitrates are absorbed by plants.

Denitrification

At various stages in the decay process, denitrifying bacteria can break down nitrates to nitrites, and then nitrites to nitrogen gas. Eventually, the nitrogen gas is released back into the atmosphere. This process

A gardening magazine stated, "grass can actually poison itself as a result of the various chemical processes that occur in the individual grass plants if the grass roots do not have enough air." To what "poison" is the magazine referring?

Teaching Tips

Discuss the two kinds of bacteria:

- Aerobic that use oxygen and anaerobic that do not.
  It is the anaerobic bacteria (denitrifying bacteria) that change the nitrates into nitrites and ultimately back to atmospheric nitrogen. This action denies the grass with the much-needed nitrates.

The magazine was referring to the lack of oxygen that allowed the anaerobes to play their role.

Bio Words

denitrification: the conversion of nitrates and nitrites to nitrogen gas, which is released into the atmosphere

phosphorous cycle: the cycling of environmental phosphorous through a long-term cycle involving rocks on the Earth's crust, and through a shorter cycle involving living organisms

is called **denitrification**, and is carried out by bacteria that do not require oxygen. Denitrification ensures the balance between soil nitrates, nitrites, and atmospheric nitrogen, and completes the nitrogen cycle.

Older lawns often have many denitrifying bacteria. The fact that denitrifying bacteria grow best where there is no oxygen may help explain why people often aerate their lawns in early spring. By exposing the denitrifying bacteria to oxygen, the breakdown of nitrates to nitrogen gas is reduced. Nitrates will then remain in the soil, and can be used by the grass to make proteins.

THE PHOSPHOROUS CYCLE

The **phosphorous cycle** is different from the water, carbon, and nitrogen cycles because phosphorous is not found in the atmosphere. Phosphorous is a necessary element in DNA, in many molecules found in living cells, and in the bones of vertebrate animals. Phosphorous tends to cycle in two ways: a long-term cycle involving the rocks of the Earth's crust, and a short-term cycle involving living organisms.

In the long cycle living things divert phosphates from the normal rock cycle. Phosphorous is found in bedrock in the form of

THE PHOSPHOROUS CYCLE

uplifting of rock
weathering of rock
phosphates in organic compounds
phosphates in rock
plants
animals
detritus
phosphates in solution
phosphates in soil
rock
precipitated (solid) phosphates
detrivores in soil

Teaching Tip

Emphasize that there are two cycles in the phosphorus cycle:

- Long term and short term.

Explain the two cycles separately and then link them together in your discussion.

Assessment Opportunity

Using the **Blackline Master 9.2: The Phosphorous Cycle,** ask the students the following:

- There are two smaller cycles in this cycle: Long term and short term. What cycle do you belong to and why?

phosphate ions combined with a variety of elements. Phosphates are soluble in water and so can be drawn out of rock as part of the water cycle. Dissolved, phosphates can be absorbed by photosynthetic organisms and so pass into food chains. Phosphates eroded from rock are also carried by water from the land to rivers, and then to the oceans. In the ocean phosphates are absorbed by algae and other plants, where they can enter food chains. Animals use phosphates to make bones and shells. When they die, these hard remains form deposits on the ocean floor. Covered with sediment, the deposits eventually become rock, ready to be brought to the surface again. The cycle can take millions of years to complete. In the short cycle, wastes from living things are recycled by decomposers, which break down wastes and dead tissue and release the phosphates. The short cycle is much more rapid.

AGRICULTURE AND THE NITROGEN AND PHOSPHOROUS CYCLES

The seeds, leaves, flowers, and fruits of plants all contain valuable nutrients, which is why we eat them. However, as crops are harvested, the valuable nitrogen and phosphorous in these plant body parts are removed and do not return to the field or orchard they came from. This diversion of nitrates and phosphate from their cycles would soon deplete the soil unless the farmer replaced the missing nutrients. **Fertilizers** are materials used to restore nutrients and increase production from land. In this activity you investigated the effect fertilizer had on the growth of algae. Some estimates suggest that fertilizers containing nitrogen and phosphates can as much as double yields of cereal crops such as wheat and barley. However, fertilizers must be used responsibly. More is not necessarily better.

Bio Words

fertilizer: a material used to provide or replace soil nutrients

The accumulation of nitrogen and phosphate fertilizers produces an environmental problem. As spring runoff carries decaying plant matter and fertilizer-rich soil to streams and then lakes, the nutrients allow aquatic plants to grow more rapidly in what is called an algal bloom. When the plants die, bacteria use oxygen from the water to decompose them. Because decomposers flourish in an environment

NOTES

A Vote for Ecology

Algae are generally thought of as simple, aquatic plants that do not have roots, stems, or leaves. A recurring problem in many bodies of water is algal bloom. An algal bloom is an abnormal increase of algae in a body of water. The most serious algal blooms are associated with human activities. Algal blooms deplete the water of oxygen and nutrients. In turn, this can kill other species in the water.

with such an abundant food source, oxygen levels in lakes drop quickly, so fish and other animals may begin to die. Dying animals can only make the problem worse, as decomposers begin to recycle the matter from the dead fish, allowing the populations of bacteria to grow even larger, and use still more oxygen.

Reflecting on the Activity and the Challenge

You have now investigated how several different types of matter cycle through ecosystems. You have also had an opportunity to learn about how humans can influence any one of these cycles. Consider how you will describe the importance of each of these cycles to the public. Also, consider whether or not the environmental issue that you have chosen deals specifically with one of these cycles. You will need to examine if any solution you provide will create a problem in any one of these cycles.

Teaching Tip

Discuss that some biologists do not consider algae as a plant. They are listed under the **Kingdom Protista** for their lack of roots, stems or leaves.

Assessment Opportunity

If the students are through with their experiment, you can give them a quiz by asking this question:

- Explain the picture on page 604, using the results of your experiment with the detergent and lawn fertilizer.

Biology to Go

1. Why is nitrogen important to organisms?
2. If plants cannot use the nitrogen in the atmosphere, how do they obtain the nitrogen they need?
3. How do animals obtain their usable nitrogen?
4. Explain why it is a good practice to aerate lawns.
5. Why is phosphorous important to living things?
6. With each harvest, nitrogen is removed from the soil. Farmers have traditionally rotated crops. Wheat, planted one year, is often followed by legumes planted the following year. Because the legumes contain nitrogen-fixing bacteria, nitrogen levels are replenished. The use of nitrogen-rich fertilizers has allowed farmers to not use crop rotation.
   a) What advantages are gained from planting wheat year after year?
   b) New strains of crops have been especially bred to take up high levels of nitrogen and harvests have increased dramatically. Speculate about some possible long-term disadvantages that these crops might present for ecosystems.
7. Before municipal sewers, the backyard outhouse was standard behind homes. They can still be found in some areas. To make an outhouse, a hole was dug in the ground to collect human wastes. Explain why the outhouse poses a risk to neighboring lakes, using information that you have gained about the nitrogen cycle.

Inquiring Further

1. The "new-tank syndrome"

Research to find out what is meant by the "new-tank" syndrome. How is it related to the nitrogen cycle?

2. Too much of a good thing

Which human activities impact on the nitrogen cycle? Choose one and explain how the impact of this activity on the environment could be reduced.

Biology to Go

1. Nitrogen is needed to make protein and also deoxyribonucleic acid (DNA).

2. Plants are helped by the nitrogen-fixing bacteria that are found in root nodules or in the soil. These bacteria change the atmospheric nitrogen into nitrates that are then absorbed by the roots of the plants. Lightning also helps because it changes the nitrogen gas into nitrates that get to the soil after they are dissolved in the rain. These nitrates are also absorbed by the roots of the plants.

3. The animals eat the plants and get the nitrogen from them.

4. Aerating the lawn deprives the anaerobic bacteria (denitrifying bacteria) to change the nitrates into nitrites. Nitrates will then remain in the soil and can be used by the grass to make proteins.

5. Phosphorous is a necessary element in DNA, in many molecules found in living cells, and in the bones of vertebrates.

6. a) An increase in wheat yield will occur.
   b) The fertilizers with their nitrogen and phosphates will eventually go into streams and lakes where they may cause algal bloom resulting in the depletion of oxygen supply causing the death of other species. The death of the other species exacerbates the problem because as the decomposers act on them, more oxygen is used, making it more unavailable to the other organisms.

Inquiring Further

1. **The "new-tank syndrome"**
   The "new-tank syndrome" talks about the lack of bacteria that can break down the ammonia present in the fish waste and other detritus like extra food into nitrites. Another set of bacteria change the nitrites into nitrates. Without these bacteria, ammonia and nitrite levels increase to dangerous amounts before they are transformed into nitrates that are relatively harmless.

   It is related with the nitrogen cycle because it also involves nitrifying bacteria that change the ammonia into nitrites that are then changed into nitrates.

2. **Too much of a good thing**
   Student answers will vary. Farmers do not rotate crops anymore. They rely on fertilizers to provide the necessary nitrates in the soil. The judicious use of fertilizers could be encouraged by giving incentives to farmers.

Teaching Tip

You may show a video on the "kaingin" (slash and burn) system of agriculture in which the farmers burn the forests to clear the land for planting purposes. When the soil becomes poor and fails to yield enough crops, the farmers go to another part of the forest to start the process all over.

Blackline Master Ecology 9.1: The Nitrogen Cycle

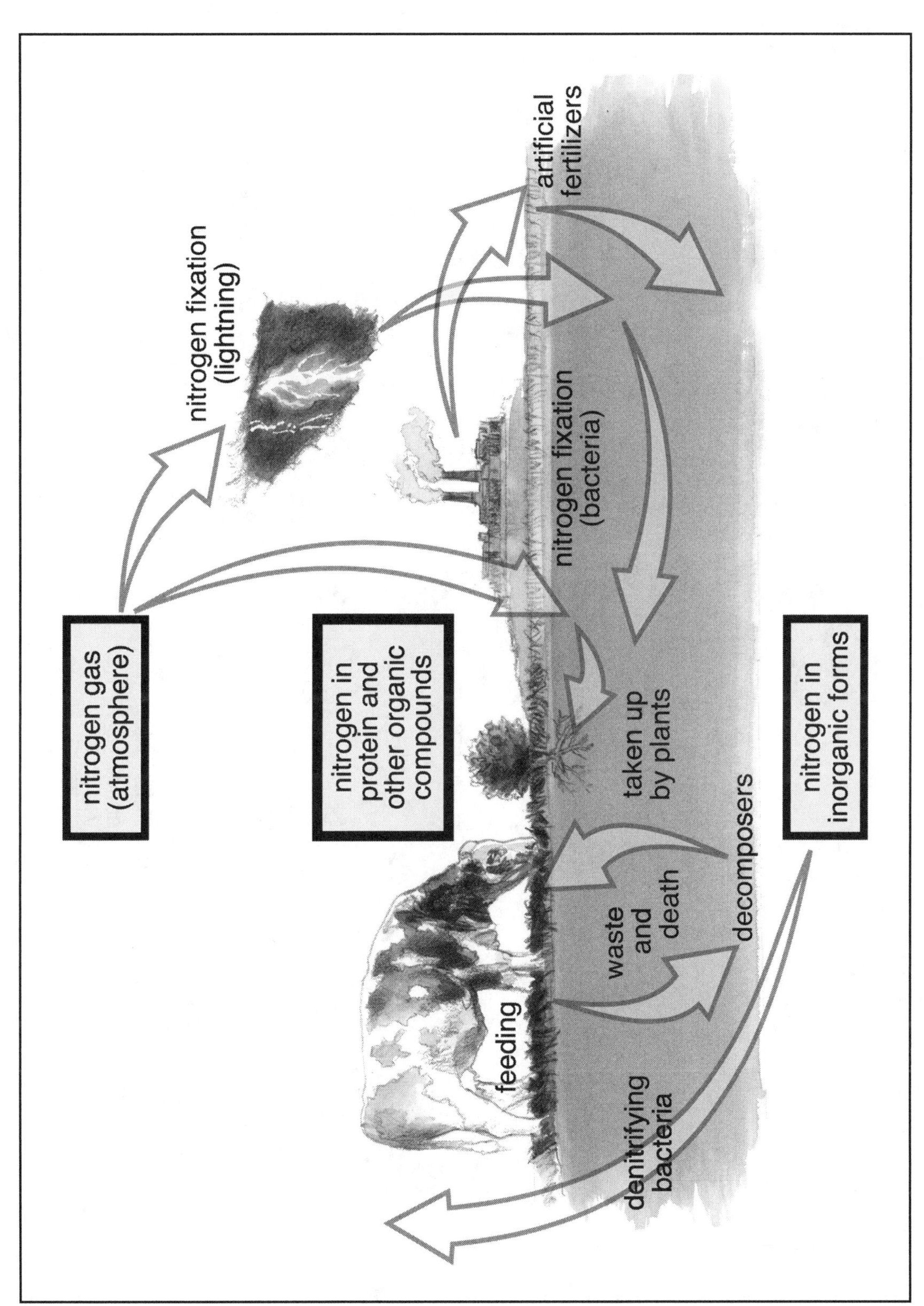

Blackline Master Ecology 9.2: The Phosphorous Cycle

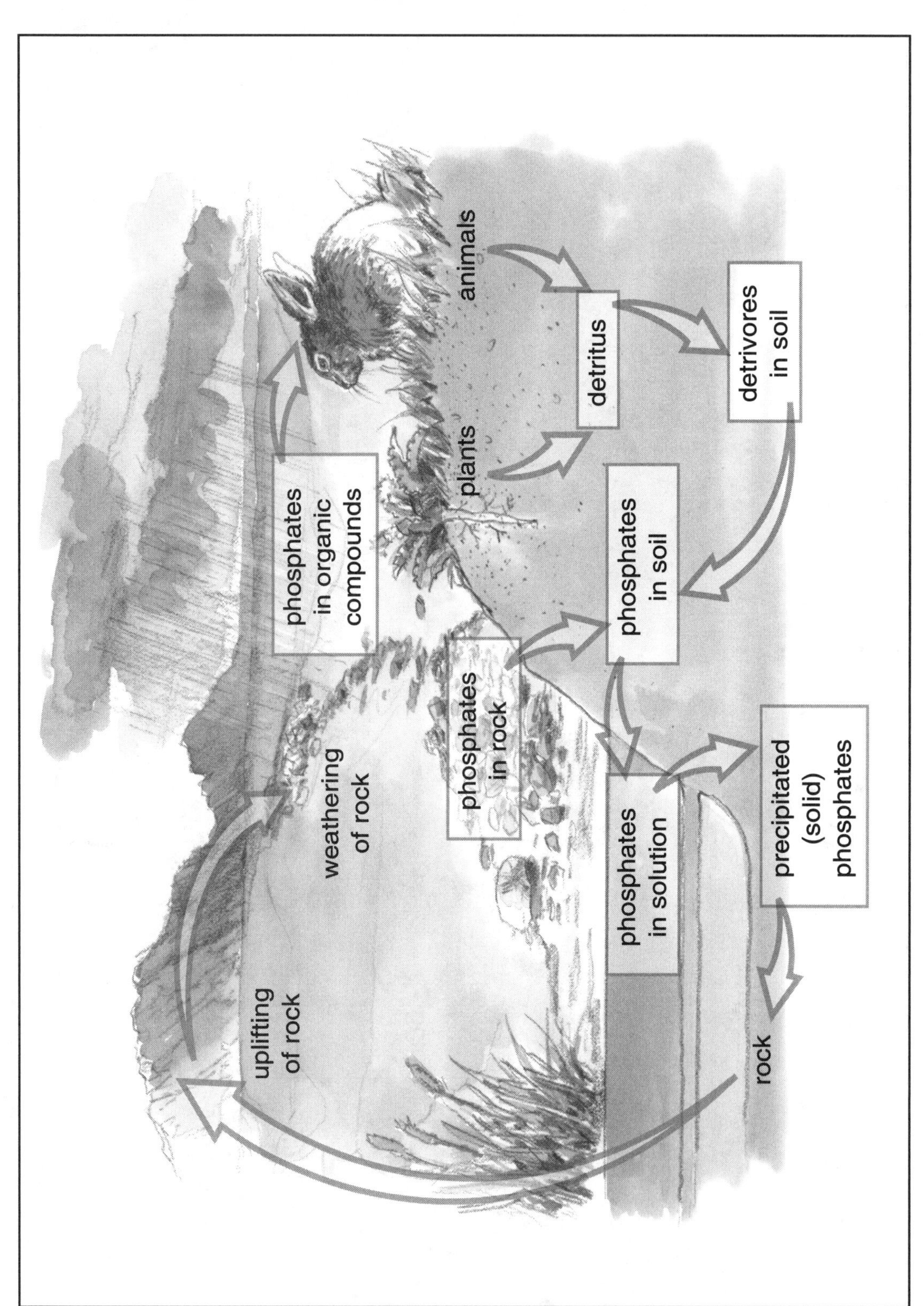

Blackline Master Ecology 9.3: Effect of Detergent and Lawn Fertilizer on Algal Growth

	Jars					
Days	A		B		C	
	Distilled water added (mL)	Observations	Distilled water added (mL)	Observations	Distilled water added (mL)	Observations
Mass of filter paper						
Mass of filter paper and algae						

Chapter 9

A Vote for Ecology

Biology at Work

Christy Todd Whitman

No person on the planet is more qualified to speak about the relationship between politics and environmental issues than Christy Todd Whitman. Whitman, 57, served as Governor of New Jersey from 1993 to 2000 and as the head of the Environmental Protection Agency from 2001 to 2003.

"Growing up on a farm I loved the outdoors. I loved to fish and swim and boat and bike and hike," Whitman says. "After seeing what happened as farms started to develop I got a real sense of the importance of protecting the environment."

Studying international government at Wheaton College in Massachusetts was also a natural extension of her curiosities as a child. "No matter what you're interested in—science or the arts or education—government impacts it in some way so I got interested in politics at an early age."

Whitman's work while EPA administrator forced her to deal with issues on a national level. She says that one of her biggest challenges was to educate the country–both individuals and corporate America–about the perils of ignoring the environment.

"A lot of the issues are very basic and we tried a lot of outreach programs. It's important to explain to people that everything you do has consequences, good and bad. For instance, we all live in a watershed so if you throw something out the window—a cigarette for example—that has an effect. If you change your oil in driveway or over-fertilize your lawn, it all will eventually wash down after a heavy rain. We found that every eight months there is as much oil deposited along the coastline from our everyday activities as was released for the Exxon Valdez spill. So what we're tying to get people to understand is that it does matter what they do. Everything is cumulative."

And that includes big business. Although, Whitman is careful to point out that economic development does not have to be mutually exclusive to environmental protection.

"When you hear environmentalists yelling about business being bad for the planet or big corporations saying that they can't be profitable and environmentally conscious–you know that neither of those are true. You can have a cleaner, healthier environment, and a thriving economy. In fact, the environment needs the money produced by a healthy economy to invest in new technology. And there isn't a country in the world or a municipality or state in the world that is going to thrive economically if their environment is not good and healthy for the people who live there." Her solution? Incentives. The theory being that if you entice industries to develop environmentally sustainable practices, everybody wins.

Regardless of the fact that she is no longer head of the nations largest environmental group or governor of the 9th most populated state in the union, Whitman is still working to educate the public in an attempt to protect the planet's natural resources. "I am very proud of programs like Energy Star, which identifies energy efficient products such as washing machines, DVD players and other technologies to consumers," Whitman says. "In 2002, purchases from Energy Star saved consumers $7 billion and greenhouse gas emissions equivalent to the removal of 15 million cars from the road." And those are the kind of environmental impact numbers every politician would brag about.

NOTES

Alternative End-of-Chapter Assessment

Multiple Choice

For numbers 1-5, choose the best answer

1. Humans have the same body symmetry as:
   a) planarians b) frogs c) earthworms d) hermit crabs
   e) all of the above

2. Decomposers can get nutrients from:
   a) plants only b) animals only c) other decomposers
   d) all of the above
   e) none of the above

3. If food chain A has three organisms and food chain B has four organisms, which of the following statements is the best answer?
   a) There was more heat lost in food chain B than in food chain A.
   b) There was more heat lost in food chain A than in food chain B.
   c) Both food chains A and B lost the same amount of heat into the environment.
   d) Both food chains A and B started with a producer.
   e) (a) and (d) are both true.

4. Which of the following makes up the growth rate?
   a) natality b) mortality c) immigration
   d) all of the above
   e) none of the above

5. Invasive species are:
   a) native species b) nonnative species c) predators
   d) never plants
   e) always animals

True or False

For numbers 6-10

6. The succession that occurs after a volcanic eruption is primary succession.
7. Plants lose water through transpiration.
8. Plants make their own food using oxygen and water.
9. Animals get their nitrogen and phosphorous from the plants.
10. Plants do both photosynthesis and respiration.

Written Response

For numbers 11-20

11. How do plants obtain the nitrogen they need when they cannot use the nitrogen in the atmosphere?

12. Explain why photosynthesis and cellular respiration go together.

13. Describe how humans can negatively impact the water cycle.

14. Differentiate primary and secondary succession.

15. What evidence indicates that there is competition taking place among the plants in a natural setting?

16. How do limiting factors play a role in the extinction of a population?

17. Explain the transfer of energy in a food chain using the laws of thermodynamics.

18. How important are the decomposers in a biological community?

19. Why is maintaining biodiversity important?

20. Using what you learned about the nitrogen cycle, explain why it is a good practice to aerate lawns.

Answers to Alternative End-of-Chapter Assessment

1. (e)
2. (d)
3. (e)
4. (d)
5. (b)
6. False
7. True
8. False
9. True
10. True

11. Plants are helped by the nitrogen-fixing bacteria that are found in root nodules or in the soil. These bacteria change the atmospheric nitrogen into nitrates that are then absorbed by the roots of the plants. Lightning also helps because it changes the nitrogen gas into nitrates that get to the soil after they are dissolved in the rain. These nitrates are also absorbed by the roots of the plants.

12. Photosynthesis releases oxygen that is needed in cellular respiration. Cellular respiration, on the other hand, releases carbon dioxide, which is needed for photosynthesis to occur.

13. Ways in which humans can have a negative effect on the water cycle are:
    a) With increasing human population, there is more demand on the freshwater resource, which remains constant. The big demand caused by the increasing population will result in very scarce freshwater resource.
    b) With increasing population, an increase in the number of industry results, which produces more domestic, agricultural and industrial wastes. These can overload water supplies with hazardous chemicals and bacteria.
    c) With the clearing of forested areas for human purposes, natural drainage patterns are replaced by road drains, sewer systems and paved land. These factors put increased pressure on the water cycle.
    d) Pollutants discharged into the air from fossil fuel burning, automobile combustion and nitrogen fertilizer processing, release sulfur and nitrous oxides. These oxides enter the atmosphere where they combine with water droplets and form acids. They return to the Earth in the form of acid precipitation.

14. The difference between primary succession and secondary succession lies in the existence or non-existence of vegetation cover before the disaster occurred. If the area was not previously covered with vegetation then it is primary succession. If the area was previously covered with vegetation, then it is secondary succession.

15. Plants would be reasonably well spaced in a natural setting.

16. If the death rate were greater than the birthrate over an extended period of time, then the population would become extinct.

17. The First Law of Thermodynamics states that energy can be transformed, but cannot be created or destroyed. Energy from the Sun is transformed by plants into a form that can be used by the consumers in a food chain. The Second Law of Thermodynamics states that in any energy transformation some energy is lost from the system. At each step of the food chain about 10% of the energy is lost.

18. The decomposers return the chemical compounds of dead organisms' bodies together with the waste products of living organisms to the environment in a form that can be used by other organisms.

19. If plants and animals are genetically similar, exposure to a disease that they are susceptible to can wipe out an entire population. With simple ecosystems being unstable, the removal of one component can result in the collapse of the whole ecosystem. And, with a lot of plants not having been identified yet, some of them may prove to be beneficial to humans.

20. Aerating the lawn deprives the anaerobic bacteria (denitrifying bacteria) to change the nitrates into nitrites. Nitrates will then remain in the soil and can be used by the grass to make proteins.

NOTES

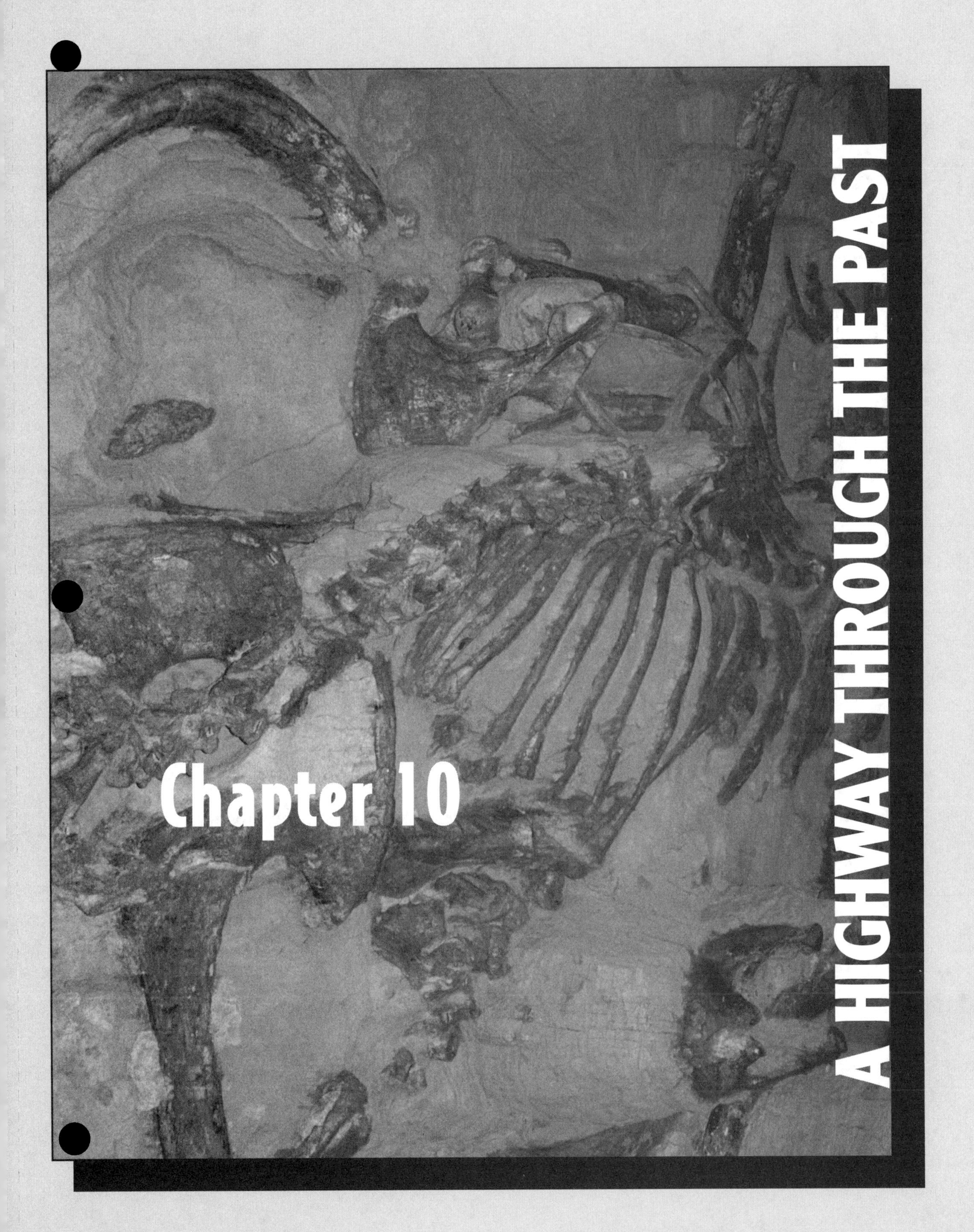
A HIGHWAY THROUGH THE PAST
Chapter 10

ACTIVE BIOLOGY
CHAPTER 10

A HIGHWAY THROUGH THE PAST

Chapter Overview

The **Chapter Challenge** is for students to represent two opposing points of view on whether the construction of a stretch of road to connect two very busy state highways should be allowed. In their arguments they should be able to indicate how the highway might impact the natural selection process that can affect the survival of a great diversity of species. They are also expected to use what they learned from the activities on the importance of fossils and how the age of fossils is determined. The activities in this chapter walk students through the process of natural selection and fossil formation.

Students begin the chapter by observing different characteristics of organisms. **Activity 1** introduces students to adaptations and their importance to the species survival. They will also observe one human characteristic and underline its importance. Next, students study how those characteristics are inherited. **Activity 2** shows students how those adapted characteristics are passed through the succeeding generation, and how the environment may influence those inherited traits. The students continue the study in **Activity 3**. This will be the fun part of the chapter. Students will simulate predation in a hypothetical environment. From the activity, students will see how unfavorable traits are eliminated by the process of natural selection. In this activity, the students are now introduced to the **Chapter Challenge. Part B** of the activity shows students a possible effect of building a four-lane divided highway on the diversity within a moth species.

Students are then introduced to fossils and how they may have ended up where they were found. In **Activity 4**, the students will enjoy modeling how fossils are made in different media. They will construct a model of fossils found in rocks, resin and ice. **Activity 5** walks students through the fossil record. Students will see the pattern of fossil deposition over time and find evidence of mass extinction and adaptive radiation.

Chapter Goals For Students

In this chapter, students will:

- Be able to explain the meaning of adaptation and how adaptation helps an organism survive in its environment
- Be able to explain how an inherited trait can be influenced by the environment
- Investigate the process of natural selection
- Model ways in which fossils are made
- Investigate fossil data for evidence of mass extinction and adaptive radiation

Chapter Timeline

The chapter may be completed in approximately 22 periods. Each activity will require two class periods to complete. You will also need a day for each activity to recapitulate. Start **Activity** 2 ahead of time. Allow time for class presentations and re-teaching. Assess student understanding at the end of the chapter. Schedule the planting of seeds on a Friday, to ensure that the 12th and 13th day of observation falls on a class day. If you want to sequentially present the activities to avoid confusion, start the planting of the tobacco seeds for **Activity 2** before you even introduce the new chapter.

Day	Activity	Homework
1	Introduce the chapter. Assign groups to respond to the Chapter Challenge. Review goals for Activity 1.	
2	Begin Part A of Activity 1. Allow 10 minutes for the 10th day observation of tobacco leaves for Activity 2. (The planting component of Activity 2 should have been started way ahead of time.)	Read the BioTalk on page 614
3	Finish Part A and Part B. Allow 10 minutes for 11th day observation of tobacco leaves for Activity 2.	
4	Recapitulation and reflection on Activity 1. Review goals for Activity 2 Allow 10 minutes for 12th day observation of tobacco leaves for Activity 2.	Biology to Go on page 618
5	13th day of observation. Answer questions to the activity.	BioTalk on page 623
6	Discuss Heredity. Recapitulation and Reflection on Activity 2. Assign groups for Activity 3.	Biology to Go on page 626 What Do You Think? on page 627
7	Part A of Activity 3.	
8	Part B of Activity 3.	BioTalk on page 632

Day	Activity	Homework
9	Recapitulation and Reflection for Activity 3. Review goals for Activity 4. Assign groups for Activity 4.	Biology to Go on page 636 Inquiring Further on page 636 What Do You Think? on page 637
10	Preparing Models.	
11	Observing Models.	BioTalk on page 641
12	Recapitulation and Reflection. Review goals for Activity 5. Assign groups for Activity 5.	Biology to Go on page 645 Inquiring Further on page 646 What Do You Think? on page 647
13	Examine Brachiopod fossils. Examine Graph A of Activity 5.	
14	Examines Graph B and C of Activity 5.	BioTalk on page 642
15	Recapitulation and Reflection.	Biology to Go on page 645 Inquiring Further on page 646
16	Recapitulation on the Chapter. Rubrics for the Chapter Challenge.	
17	Preparation time to respond to the Chapter Challenge.	
18	Preparation time to respond to the Chapter Challenge.	
19	Presentations.	
20	Presentations.	
21	Review.	
22	Assessment.	

National Science Education Standards

Chapter Challenge

The building of a road to link two major highways provides the **Scenario** for this chapter. Students are challenged to present different arguments for and against building of the highway using principles they learned from this chapter in a town hall meeting setting.

Chapter Summary

To gain understanding of science principles to meet this challenge, students work collaboratively on activities to learn how changes in the environment might affect the diversity of species in a particular area. They are to explore what happen to species in the past to gain better understanding of the concepts involved.

CONTENT STANDARDS

Unifying Concepts and Processes

- Systems, order, and organization
- Evidence, models, and explanation
- Change, constancy, and measurement
- Form and function

Science as Inquiry

- Develop abilities necessary to do scientific inquiry
- Identify questions and concepts that guide scientific investigations
- Design and conduct scientific investigations
- Use technology and mathematics to improve investigations and communications
- Formulate and revise scientific explanations and models using logic and evidence
- Recognize and analyze alternative explanations and models
- Communicate and defend a scientific argument
- Develop understandings about scientific inquiry

Life Science

- Diversity and adaptation of organisms
- Biological evolution

Science in Personal and Social Perspectives

- Science and technology in local, and global challenges

Key Science Concepts and Skills	
Activities Summaries	**Biology Principles**
Activity 1: Adaptations Students begin the chapter by looking at photographs of different types of adaptations. These observations help them understand the importance of adaptation to a species survival. The student will also learn two types of adaptation; structural and behavioral.	• Adaptation • Structural and behavioral adaptations • Body symmetry • Plant and Animal adaptations
Activity 2: Is It Heredity or the Environment? To understand the influence of the environment to the inherited traits, the students will observe how the absence of light might affect the color of the leaf of a tobacco plant. The reading explains how traits are transmitted from parents to offspring.	• Heredity • Law of Dominance • Genotype and phenotype
Activity 3: Natural Selection Students simulate predation in a particular environment. They will also study the effect of a four-lane highway to a moth population. The students will learn how the process of natural selection occurs from this activity.	• Natural Selection and evolution • Theory and hypothesis
Activity 4: The Fossil Record The students will model how fossils are formed. The reading explains the different types of fossils and how some fossils are used as index to divide the fossil records into geologic era.	• Fossilization • Body, index, trace fossils • Fossil Record • Geologic time periods
Activity 5: Mass Extinction and Fossil Records The students will analyze graphs of fossil data to identify periods of mass extinction and adaptive radiation.	• Mass Extinction • Adaptive Radiation • Geologic time periods • Ecological Niches

Equipment List for Chapter 10

Materials needed for each group per activity.

Activity 1

- masking tape
- 5 coins

Activity 2

- blotting paper
- 4 Petri dishes
- tobacco seeds
- black plastic bags or lightproof containers

Activity 3

- a full sheet of newspaper
- 50 half-inch squares cutout of newspapers
- 50 half-inch squares cutout of white papers
- 50 half-inch squares cutout of red papers

Activity 4

Station 1

- plaster (a cupful)
- 1 clamshell
- petroleum jelly
- confetti
- 1 cup
- masking tape
- small hammer

Station 2

- 1 small paper plate
- hot-glue gun with glue sticks
- tweezers
- small seeds

Station 3

- 1 paper cup half-full of water per group
- 1 small seed, shell, or any other small object

Station 4

- modeling clay
- rolling pin or bottle
- plaster
- petroleum jelly

Activity 5

No materials are needed for **Activity 5**

A Highway Through The Past

Scenario

After much study, the State Department of Transportation decided that a stretch of road was needed to connect two very busy state highways. The study had included environmental assessment of the area that the road would be covering. Much to the frustration of the local people, this study had taken over a year to complete.

Now, by state law, before any road construction could begin paleontologists (scientists who study past life) from the state university had been given six months to study the land that would be covered by the new road. The local residents who would use this new road were very upset because of this further delay of six months. They were even more upset when they found out that the findings of the paleontologists would delay the road construction longer than six months, and maybe even indefinitely.

Chapter Challenge

The State Department of Transportation has concluded that in order to allow area residents a chance to express their views, a town-hall meeting should be held to discuss the issue. At this meeting, the paleontologists would also be given an opportunity to provide their findings and explain why they have asked for a delay or even an indefinite postponement of the construction. For your challenge, you may be asked to represent someone who is against building the road, someone who is for building the road, or one of the group of paleontologists. Or, you may be asked to represent a member of one of the levels of government involved in this project to explain why it is necessary to have all these studies

608

Scenario and Chapter Challenge

The **Chapter Challenge** involves a town hall meeting. Review with students how town hall meetings are presented. Explain to the students that all the groups in a town meeting will be provided an equal chance to present their arguments. Reserve time for questions and answers after all sides have presented their argument.

and subsequent delays. Your teacher will act as the chairperson. As you prepare for and participate in role-playing a town meeting, you will be expected to:

- explain why fossils are important
- describe how the age of a fossil can be determined
- indicate how a highway might impact on the natural-selection process of the living organisms in the area
- explain why a great diversity of species is important for the survival of a community.

Criteria

How will your performance at the meeting be graded? Keep in mind that not everyone will be arguing every side of the issue. It is very important that you decide before the work begins how each person will be graded. Discuss this with your small group and then with your class. You may decide some or all of the following qualities are important:

- completeness and accuracy of the science principles presented in your side of the argument
- accuracy of the science principles used to dispute your opponents' positions
- forcefulness or conviction with which you present your argument
- quality of questions you pose to the government officials.

Once you have determined the criteria that you wish to use, you will need to decide on how many points should be given to each criterion. Your teacher may wish to provide you with a sample rubric to help you get started.

Criteria

It is important that students claim ownership of their assessment. One way of involving students in developing their own assessment is to come up with a rubric for assessing the **Chapter Challenge.**

You can elicit suggestions from students on what should be included in the rubric in a whole-class setting. This will take a lot of time. One method to maximize student participation is asking the groups to come up with a criteria chart. After they are done with their criteria chart, convene a group of representatives from each group to narrow down the criteria to a certain number. Present the criteria to the class for a final discussion.

Suggested Assessment Rubric for Chapter Challenge

Meets the standard of excellence. **5**	• All four of the criteria set for the challenge were met. • Arguments presented are supported by data and very convincing. • Additional research beyond basic concepts presented in the chapter is evident. • The presentation is interesting and all members of the group participated.
Approaches the standard of excellence. **4**	• Three of the criteria set for the challenge were met. • Arguments presented are supported by data. • The presentation show creativity and all members of the group participated.
Meets an acceptable standard. **3**	• Two of the criteria set for the challenge were met. • Arguments presented are reasonable. • All members of the group participated.
Below acceptable standard and requires remedial help. **2**	• One of the criteria set for the challenge was met. • Arguments presented have no basis. • Some members of the group participated.
Basic level that requires remedial help or demonstrates a lack of effort. **1**	• None of the criteria set for the challenge was met. • Arguments presented have no basis. • Some members of the group participated.

NOTES

ACTIVITY 1– ADAPTATIONS

Background Information

Adaptation

In **Chapter 9**, the students were introduced to the diversity of living organisms. Organisms exist in a variety of forms and live in a lot of different environmental conditions. They live, interact and compete with each other. Each organism has particular characteristics adapted to its way of life.

All organisms require energy to support their metabolic processes. Competition for space and source of energy often define the relationship and interaction of different species. All organisms are part of the food chain. They are either the predator or the prey of some species. Organisms compete for a limited source of energy and resources.

The struggle for survival puts a stress on the organism's ability to propagate and multiply. Organisms that are able to survive the food chain and the environment often are able to reproduce and propagate their species. Those who do not survive will eventually become extinct.

Adaptation is an organism's ability to adjust to changes in its environment and compete and interact with other organisms. Organisms developed various adaptations that enable their species to survive. Species with a wider variety of characteristics have a better chance of surviving changes in the environment.

Both plants and animals employ a variety of adapted characteristics that best suit their environment. This adaptation can be behavioral or structural.

Structural adaptations are body parts of an organism that are suitable for its way of life. Moles have pointed claws that are adapted for digging. This enables the mole to burrow underground. The cheetah has well developed limbs that enable it to run fast to catch its prey. Not all animals are adapted to run fast to catch its prey or escape being caught. They use other adaptations to be equally successful. Some organisms use camouflage to hide from predators or stalk their prey. Other animals use mimicry to fool predators and increase their chances from being eaten.

Organisms also developed behavioral adaptations. Most organisms in extreme conditions lay dormant in less favorable conditions and active when the climatic conditions are favorable. Some plants and animals time their reproduction on the changes in the season to increase the chances for survival of their offsprings. Plants scatter their seeds in summer or fall, lay dormant during the winter and then grow during spring. Desert frogs take advantage of pools that develop during brief storms, lay their young, then hibernate when the pools start to run dry. A cacophony of mating calls often follows the brief desert storm.

Plants and animals that are successfully adapted to their environment are able to reproduce and pass their characteristics to the succeeding generations.

The focus of this chapter of *Coordinated Science for the 21st Century* is to make students aware that organisms have various characteristics adapted to its way of life. These adaptations help the organism survive its environment. The activity shows students a variety of adaptations of plants and animals. This activity will enable students to see organisms have develop different methods of adaptation that help them survive.

Part A of the activity exposes students to some structural adaptations of plants and animals. An example on how adaptation may help propagate the species is also provided.

Part A of the activity also shows students that not all adaptations are structural. Examples of behavioral adaptations that help plants and animals survive are included.

Part B of the activity also examines human structural characteristics that most people take for granted. The opposable thumb, a characteristic we share with primates, enables dexterity of movement of our hands.

NOTES

Goals and Assessments

Goal	Location in Activity	Assessment Opportunity
Observe a group of diverse organisms.	**For You To Do** Part A	**Biology to Go** Question 1.
Speculate how adaptations help an organism survive in their environment.	**For You To Do** Parts A and B	**Biology to Go** Question 3.
Distinguish between structural and behavioral adaptations.	**For You To Do** Part A	**Biology to Go** Question 2.

Activity Overview

This activity shows how different organisms adapt to their environment. Students will look at pictures of different organisms in their natural environment. They will see that plants and animals employ various adaptive mechanisms in their daily existence. The students will compare how the different structural makeup of these organisms is advantageous to their survival as they interact with their environment. The students will also be able to see that not all adaptations are structural. They will see that plant or animal behaviors increase their ability to adapt to changes in the environment.

Students will also examine one of their own structural adaptations, the opposable thumb.

Preparation and Materials Needed

Preparation

You may ask the students the day before the activity to come to school wearing buttoned shirts and laced shoes on the day of the activity. This will be needed for **Part B** of the activity.

No other preparation is required for this activity.

Materials/Equipment Needed (per group)

- masking tape
- coins

Learning Strategies for Students with Limited English Proficiency

1. Point out new vocabulary in context. Practice using the words as much as possible.

camouflage	behavioral adaptation	fleshy
structural adaptation	mimicry	opposable thumb

   Provide all students with an opportunity to observe the living organisms using primarily their natural curiosity and observational skills. As the students make their observations, encourage them to connect what they see to the new vocabulary words. Keep in mind that the objective of the activity is to sharpen students' observational skills and to gain an appreciation for the diversity of life. The vocabulary is there to guide students in their observations. Substitute vocabulary words or use diagrams, as appropriate for your class.

2. As in previous activities consider using a CLOZE activity at the end of this activity. Ask students to describe their own behavioral and structural adaptations. Ask the students to explain the importance of those adaptations to their existence.

NOTES

Activity 1 Adaptations

GOALS

In this activity you will:

- Explain the meaning of adaptation.
- Speculate how adaptations help an organism survive in their environment.
- Distinguish between structural and behavioral adaptations.

What Do You Think?

Imagine surviving a temperature of −50°C and a blinding snowstorm. Imagine surviving a temperature of 50°C in an extremely dry landscape.

- **How are plants and animals that live daily in these environments adapted for survival?**

Write your answer to this question in your *Active Biology* log. Be prepared to discuss your ideas with your small group and other members of your class.

For You To Do

Part A: Observing Adaptations

An adaptation is an inherited trait or set of traits that improve the chances of survival and reproduction of organisms. In this part of the activity you will look at photographs of animals to

What Do You Think?

- Organisms living in extreme conditions have body structures adapted to that environment. They have also developed behavioral adaptations to increase their chances of survival.

In extremely cold environments animals either have thick specialized furs or insulating fat underneath their skin. Some organisms hibernate or slow down their metabolism during the winter season. Plants in cold regions are normally low and grow slowly. They lie dormant during the winter and flower and grow during the brief summer.

In extremely hot conditions, animals are lean and are adapted to conserve water. Others get water through their metabolic processes. These animals feed at night when it is a lot cooler. Plants develop fleshy stems to hold water and small leaves to minimize evaporation through transpiration.

Student Conceptions

Students may be able to see the advantage of adaptation to the individual organism but may not see the connection to the survival of the species. Discussion on how adaptation benefits the species as a whole, and how those adaptation are passed to future generations may clarify this misconception.

observe and speculate about how the different types of adaptations help the organism survive.

1. Look closely at the following photographs. There is a living organism in each picture.
   a) Which organisms are exhibiting camouflage?
   b) How could this adaptation help an organism in capturing prey?
   c) How could this adaptation help protect the organism from predators?
   d) What other animals can you think of that use this type of adaptation for protection?
2. Some animals are not adapted to disappear into the background, but rather stand out.

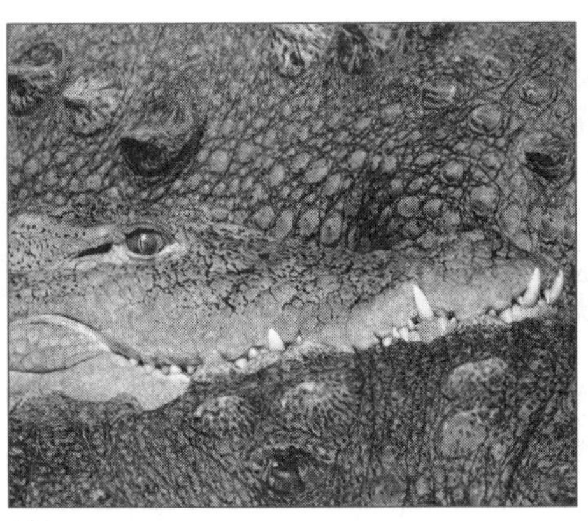

Alligator.

Praying Mantis.

Snowshoe hare.

Chameleon.

For You To Do

Teaching Suggestions and Sample Answers

Teaching Tip

Before the start of the activity, go over **Activity 1** of **Chapter 9**. It may help to discuss the relationship of body structures to its specific function and how those body structures are advantageous to the individual.

Part A: Observing Animal Diversity

1. a) All the pictured organisms are exhibiting camouflage.
   b) These organisms can remain unseen by their prey.
   c) They disappear in the background.
   d) The polar bear in the snow or the lion in the savanna.

A Highway Through the Past

Hawk moth.

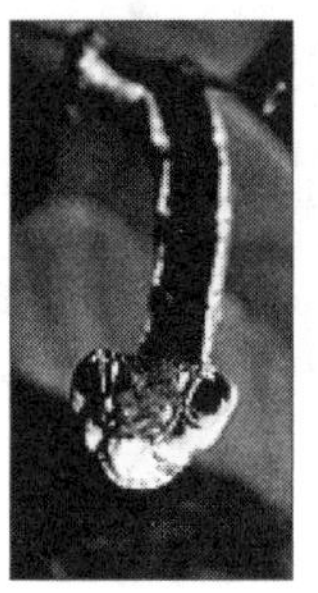

Hawk-moth caterpillar.

Monarch butterfly.

Viceroy butterfly.

Look at the photographs of the hawk moth and caterpillar.

a) At first glance, of what animal does each remind you?

b) Why would birds avoid an animal with large eyes at the front?

c) What advantage does this adaptation present for the moth and caterpillar?

3. A monarch butterfly stores bad-tasting chemicals in its body that birds hate. The viceroy butterfly also has a bitter taste.

   a) The monarch butterfly is brightly colored. Why do you think that this would be an advantage for the monarch butterfly?

   b) Would the bright colors and bitter taste protect all monarch butterflies? Explain your answer.

   c) Compare the appearance of the monarch and viceroy butterflies. Can you distinguish between them?

   d) How would the viceroy butterfly's coloration be an advantage for its survival?

4. Adaptations are not limited to animals. Look closely at the plants or plant parts shown on the next page for their adaptations to the environment.

2. a) The hawk moth looks like the eyes of a hawk, and the caterpillar looks like a snake.
   b) Large eyes are usually associated with large predators.
   c) The moth and the caterpillar can be mistaken as predators.

3. a) Predators will avoid the monarch butterfly because of its bitter taste.
   b) No. Other predators may be able to tolerate the bitter taste.
   c) At first glance, the Monarch butterfly and the Viceroy butterfly look the same.
   d) The Viceroy butterfly can be mistaken for a Monarch butterfly. Predators will avoid them.

Teaching Tip

Students often fail to see the advantage of mimicry in relation to the Monarch butterfly. You may try to clarify by asking students "If you do not like spinach, will you eat anything that looks like spinach even if they say it is good?"

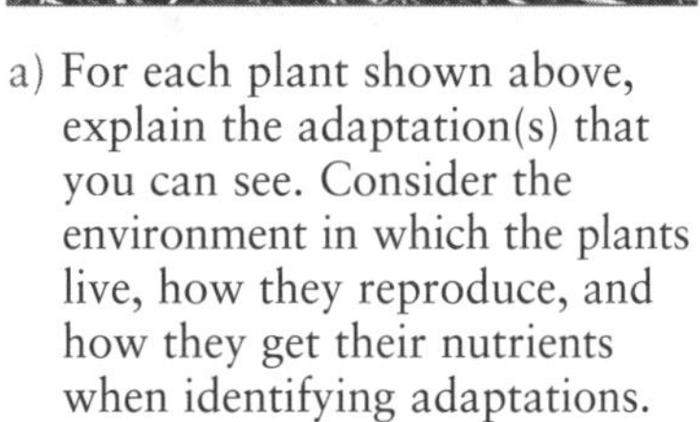

a) For each plant shown above, explain the adaptation(s) that you can see. Consider the environment in which the plants live, how they reproduce, and how they get their nutrients when identifying adaptations.

5. Not all adaptations need to be structural. Some adaptations can be behavioral.

   a) How is each animal in the photographs adapted to a change in the environmental conditions from summer to winter?

   b) How do other animals adapt to an environmental change? Give at least two examples.

   c) What type of behavioral adaptation is the plant at right exhibiting?

4. a) The fleshy stem of the cactus stores water.

   Bristles protect the seeds but can easily be dispersed by the wind during the brief summer.

   Showy flowers attract insects and help spread pollen to other flowers.

   Carrots have fleshy roots to store food.

   The Venus flytrap traps insects for their nutrients.

5. Student answers will vary.
   a) Birds migrate to warmer climates during the winter. Bears hibernate during the winter.
   b) A dog pants when it is hot, and birds spread their feathers to better trap still air during cold temperature.
   c) The plant is following the light source.

6. Invent an organism with specific adaptations. Consider one of the following:
   - camouflage
   - mimicry
   - warning coloration

Part B: How Well Adapted Are You?

In this part of the activity you will have an opportunity to examine one of your own adaptations that you probably take for granted.

1. Using masking tape, have your partner tape your thumb to your index finger on each hand. After your thumbs are securely taped, try each of the following activities. Rank the difficulty of each activity on a scale of 1 to 5.
   - picking up and carrying your textbook;
   - writing your name and address on a piece of paper;
   - picking up five coins from the floor and placing them in your pocket;
   - unbuttoning and buttoning a button;
   - tying up a shoe.

   a) Did you find any of the activities impossible?

   b) How did your ratings compare with others in your group and in your class?

   c) Why do you think that an opposable thumb is an important adaptation for humans? (An opposable thumb is an arrangement in which the fleshy tip of the thumb can touch the fleshy tip of all the fingers.)

   d) Do any other animals have opposable thumbs?

BioTalk

Adaptation

Diversity is a striking feature of living organisms. There are countless types of organisms on Earth. They are the result of repeated formation of new **species** and adaptation. There is a type of organism that can live in almost every type of environment on Earth. Living organisms are unique in their ability to adapt. The accumulation of characteristics that improve a species' ability to survive and reproduce is called **adaptation**. Adaptation occurs over long time periods. It is the environment that "selects" the best and most useful inherited variations. In this activity you observed just a few of the large number of adaptations that exist.

Bio Words

species: a group of organisms that can interbreed under natural conditions and produce fertile offspring

adaptation: an inherited trait or set of traits that improve the chance of survival and reproduction of an organism

6. Student answers will vary.

   Camouflage – an organism that can change color according to the environmental changes in the season.

   Mimicry – an organism that looks like a leaf or a rock in its habitat.

   Warning coloration – an organism that looks like a predator further up in the food chain.

Teaching Tip

Students may have difficulty inventing their own organism. Direct students to go over **Steps 1-3** in the activity to give them a clue on how their organism should look.

Part B: How Well Adapted Are You?

1. The ratings will vary depending on the individual's dexterity. Answers to **Questions a)** and **b)** will also vary.

Teaching Tip

You may have to suggest how the students should rank the difficulty for purposes of uniformity of scale. One possible ranking is as follows:

5 – very hard
4 – hard
3 – normal
2 – easy
1 – very easy

Have the class put their rating by groups on the board for class comparison.

   c) An opposable thumb enables a human to grasp and hold objects.

   d) Other primates.

Teaching Tip

Activity 1 is important because it provides the basis for the succeeding activities in this chapter. This may be the time to make connections with the lessons in **Chapter 9**, particularly **Activities 2, 3, 4,** and **5.** This will be valuable in understanding the next activities.

Animals Adapt to the Demands of Their Environments

Animals cannot make their own food. Therefore, they must usually seek food. As a result, adaptations that allow animals to move are favorable. Movement is easier if the organism is elongated in the direction of movement. Fish, for example, are streamlined. This reduces water resistance as they swim. It is also easier to move if the sensory organs are concentrated in the head. The organs that detect food, light, and other stimuli should be in a position to meet the environment first. An organism can move more easily if it has a balanced body.

Animals have the type of body plan that is best suited to their lifestyle. The symmetry of an organism gives clues to its complexity and evolutionary development. Higher animals, including humans, are symmetrical along the mid-sagittal plane. This body plan is referred to as **bilateral symmetry**, in which the right and left halves of the organism are mirror images of each other. Some animals, however, are **radially symmetric**, or symmetric about a central axis.

How is body symmetry related to the speed at which an animal moves and to brain development? In general, animals that display radial symmetry are not highly adapted for movement. One explanation for the slower movement can be traced to the fact that no one region always leads. Only bilaterally symmetrical animals have a true head region. Because the head, or anterior region, always enters a new environment first, nerve cells tend to concentrate in this area. The concentration of

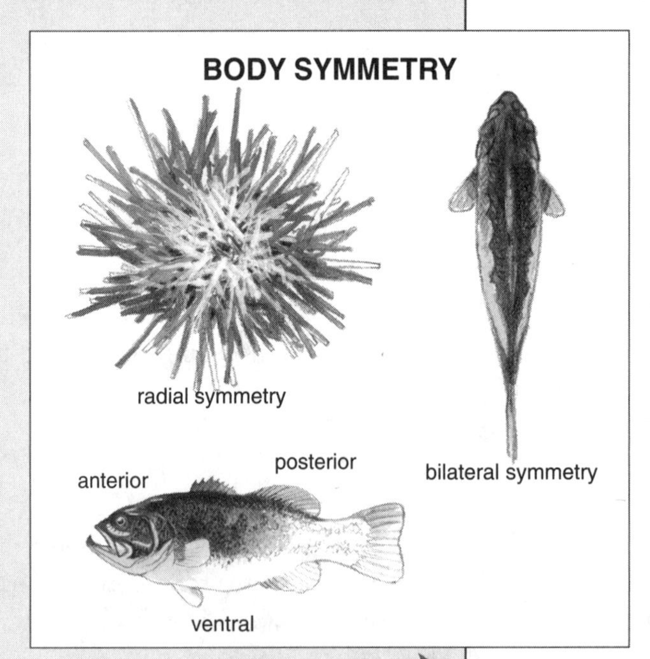

Bio Words

bilateral symmetry: a body plan that divides the body into symmetrical left and right halves

radial symmetry: a body plan that is symmetrical about a center axis

BioTalk

Teaching Tip

Students often do not realize that plants also exhibit adaptation. They do not realize that plants also employ various methods like animals do to get the resources they need and adjust to the various environmental conditions. In the study of life science, there is an over-emphasis on animals, and plants are often neglected. This may be an opportunity to showcase the plants. Students may be assigned to do a research project on how different plants adapt to their environment.

Assessment Opportunity/Teaching Tip

You may ask students to prepare two tables with two columns each. One table is for structural adaptation and the other is for behavioral adaptation. One column for **Animal Adaptations** and the other column for **Plant Adaptations**. Ask the students to list as many adaptations that were mentioned in the reading.

A Highway Through the Past

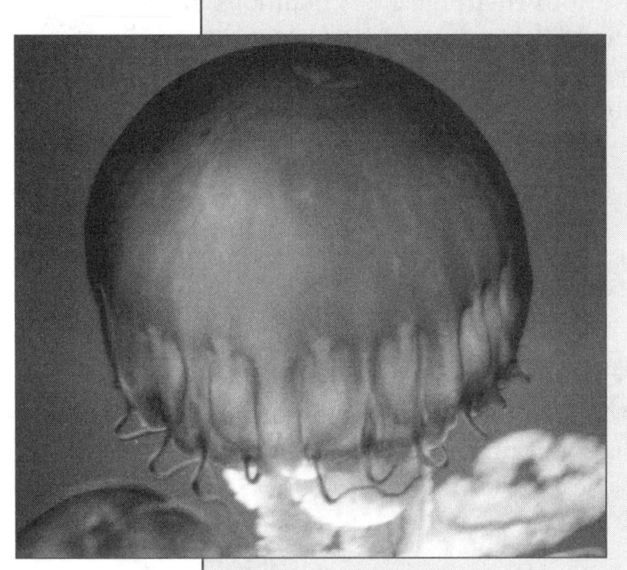

The jellyfish is a group of structurally simple marine organisms. The jellyfish has no head and a nervous system without a brain. The body exhibits radial symmetry.

nerve tissue at the anterior end of an animal's body, is an adaptation that enables the rapid processing of stimuli such as food or danger. Not surprisingly, the faster the animal moves, the more important is the immediate processing of environmental information. Every environment places special demands on the organisms living there. Seawater is fairly uniform. It poses the least stress for animal life. Oxygen is usually adequate. The temperatures and salt content are fairly constant. There is little danger that the organism will dry up. In contrast, the salt and oxygen contents of fresh water vary greatly.

Organisms that live in water have special adaptations. Gills, for example, allow the organisms to use the oxygen found in water. On land, oxygen is plentiful. However, the organisms that live there must protect themselves from the dangers of drying up. These dangers increase greatly because air temperatures change daily and seasonally. Air does not provide the same buoyancy as water. Therefore, large terrestrial, or land-dwelling, animals require good supportive structures. On the other hand, there is less resistance to movement in air than in water. Arms and legs, which would hinder an animal's movement in water, may help on land. Thus, long appendages specialized for locomotion have evolved in terrestrial animals.

Plant Adaptations

Plants lack the ability to move and must survive in the environment in which they are living. A plant must do more than simply survive and grow bigger. It must grow in such a way that it can take the best advantage of the light, water, and other conditions available to it.

NOTES

Desert plants are an excellent example of adaptation to an environment. Some have a thick waxy coating to prevent them from drying out. Some have long vertical roots enabling a plant to reach water sources beneath the soil. Others develop shallow roots that extend horizontally. This maximizes water absorption at the surface. Many desert plants have small and narrow leaves. This decreases the heating from the Sun.

Even though plants are not able to move, they are still able to disperse. They produce seeds and fruits or other reproductive structures that may be distributed far from the parent plant.

Some plant adaptations are also behavioral. A vine spreads its leaves outward and receives as much light energy as possible. It sends its roots downward and receives more water. Tendrils of a vine touch an object and quickly coil it. This secures the vine in its upward growth. A vine would not live very long if it did not send its roots downward and its stem upward. The manner of plant growth is believed to be governed chiefly by hormones that are produced within the plant. The hormones are produced in response to conditions around the plant such as sunlight and gravity. Thus, the plant can fit itself to the environment in which it lives.

Tendrils are modified stems or leaves that wrap around a support. They enable the plant to achieve fairly extensive horizontal and vertical spread without the use of much energy, since they don't have to support their own weight. Tendrils seem to respond to touch so if the stem or tendril touches an object, it wraps around it. This response is known as thigmotropism.

Some plants have even become adapted to feeding on animals. In this activity you looked at the Venus flytrap. Its leaves have been adapted to capture prey. These plants do photosynthesize. However, these plants live in bogs where there is very little nitrogen available. Therefore, they require the nutrients they receive from digesting their prey. Of course, the plant must therefore also be adapted to digest its prey with the secretion of chemicals.

NOTES

Reflecting on the Activity and the Challenge

In this activity you had an opportunity to look at adaptations of different organisms. You learned that every environment places various demands on the organisms living there. Organisms have developed special adaptations for living in any given environment. The animals and plants in the area of the highway construction have also adapted to their environment. In an environmental study scientists would have assessed the impact the highway would have had on the animals and plants. You may need to address this issue in the town-hall meeting if you are representing a government employee.

1. Explain the term adaptation.
2. Distinguish between a structural and a behavioral adaptation.
3. a) How can an animal's structure help it survive in different environments? Give three examples.

   b) How can an animal's behavior help it survive in different environments? Give three examples.
4. Do all animals living in the same environment have similar adaptations? Explain your answer.
5. A cross section represents a cut through the middle of an animal's body. Below are cross sections through an earthworm, sand worm, and a primitive insect.

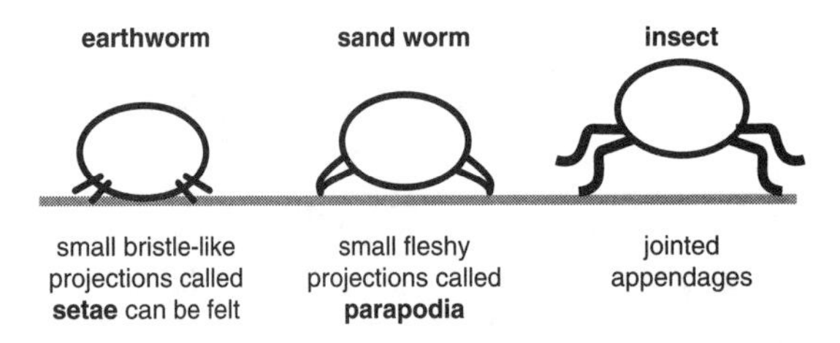

   a) The jointed appendages of the insect lift the body from the ground. How does this help the insect move?

   b) What advantages might the fleshy projections of the sand worm have over the bristle-like projections of the earthworm?

   c) Predict which animal would be the fastest and give your reasons.

Biology to Go

1. An inherited trait or set of traits that improve the chance of survival and reproduction of the species.
2. Structural adaptations are body structures that are advantageous to an organism's chances for survival. Behavioral adaptation is what an organism does to better its chances for survival.
3. Students' examples will vary.
   a) 1. Fishes have gills to filter oxygen from the water.
      2. Arctic animals have thick furs and insulating layers of fat.
      3. Birds wings are adapted for flying.
   b) 1. Desert animals hunt at night when it is cooler.
      2. Bears hibernate during the winter.
      3. Migrating birds fly to warmer regions during the winter.
4. Animals may develop similar adaptations as well as adaptations unique to the species depending on the traits they inherited from their ancestors. As shown in the activity, different species use camouflage, while others use mimicry or warning coloration in the same environment. Others develop unique behavioral adaptations.
5. a) This reduces contact with the ground to facilitate faster movement.
   b) They provide better contact with the loose sand than the bristles of the earthworm.
   c) The insect would be the fastest because the body is lifted away from the ground. This will lessen drag caused by the contact with the ground.

Inquiring Further

1. Animal adaptations to the arctic

Keeping warm is no easy task in the arctic where frigid weather lasts almost nine months of the year and where temperatures can plunge to −55°C. Even during the brief summer, when the land thaws and the Sun never sets, a sudden snowstorm can freeze everything. What adaptation have animals that live in this region developed?

2. Animal adaptations to the desert

Lack of water creates a survival problem for all desert organisms. However, animals have an additional problem. The biological processes of animal tissue can function only within a relatively narrow temperature range. Fortunately, most desert animals have evolved both behavioral and structural adaptations. Research the adaptations of animals living in desert regions.

Seals are well adapted to a cold environment. Their slick fur sheds water, and a thick layer of blubber beneath the skin keeps them warm in frigid temperatures.

The desert tortoise retreats to its burrow during the hottest times of the summer days. In the cold of winter it hibernates in its underground burrow.

Inquiring Further

1. Animal adaptations to the arctic

The students will not have any difficulty finding the information on animal adaptations in a very cold environment, but may have difficulty finding plant adaptations in the arctic region. Provide extra incentives for students who are able to discuss various structural adaptations of plants in the arctic.

2. Animal adaptations to the desert

There are various examples of adaptation in a desert environment. For plants, most of the examples the students will find are structural and for animals most of the examples are behavioral. Provide extra incentives for students researching on behavioral adaptations of plants and structural adaptations of animals.

ACTIVITY 2– IS IT HEREDITY OR THE ENVIRONMENT?

Background Information

Genetics

The study of genetics, the branch of Biology that deals with the study of inheritance, started with the experiments conducted by Gregor Mendel on pea plants. He was able to explain how characteristics are passed from one generation to the next generation. Modern genetics expanded his findings as technology became better in determining how traits are expressed and inherited.

The characteristics or traits of an organism are passed from parents to offspring through the chromosomes that are present in the nucleus of cells. This information is contained in the two sets of chromosomes of the individual in the form of deoxyribonucleic acid or DNA. During reproduction each parent contributes one set of chromosomes to the offspring.

There are two copies of a gene called alleles or factors. Factors can either be dominant or recessive. The recessive gene is usually masked unless both parents contribute the recessive genes. This explains why certain characteristics may show up on offsprings even if the trait does not seem to be present in their parents.

Organisms carrying both dominant factors or both recessive factors are known to be Pure, and organisms carrying one dominant factor and a recessive factor are known to be Heterozygous. The dominant trait will show in a heterozygous individual.

Some characteristics are multiple allele traits, or have more than two versions of a trait. The human blood type ABO is an example of multiple allele traits. This human blood type has two dominant traits, type A and type B, and a recessive trait, which is type O. If the offspring inherits the allele or factor A from one parent and the allele B from the other parent, the offspring is type AB, a codominant trait.

In most instances, traits are polygenic. Several genes rather than just one gene control the expression of the trait. This explains why certain characteristics have a wide range of variations.

The environment may also influence individual characteristics. These conditions, however, may exist temporarily depending on the changes in the environment. These characteristics revert back to its original form when the conditions return to normal

Plant Germination and Photosynthesis

Plants make their own food through a process known as photosynthesis. They convert light energy and store it as food in various parts of the plant. The parts of plants where food is stored that students are familiar with are root crops, fleshy stems, and fruits. When a plant needs the energy, the stored food is converted back into energy in the form of Adenosine Triphosphate or ATP. ATP is known as the energy currency of the cell.

The chlorophyll found in the chloroplast of the plant leaves is responsible for photosynthesis. They provide the green color of the leaf. Depending on light intensity, the leaf color can be green, red or yellow. Plants with red or yellow leaves are adapted for locations where light intensity is low. This explains why leaves from deciduous trees change color in winter. Plants lacking any pigmentation are displaying a characteristic called albinism.

Seeds contain stored food that it will use for germination. Most plants just require water and the right environmental conditions to start germination. Since they have stored energy, they do not need sunlight for germination. The seeds have enough stored energy to support it through germination. In this activity, the students will examine how a particular trait's expression can be influenced by the environment.

Goals and Assessments

Goal	Location in Activity	Assessment Opportunity
Observe how an inherited trait can be influenced by the environment.	For You To Do	Biology to Go Questions 5 and 7.
Distinguish between phenotype and genotype.	BioTalk	Biology to Go Questions 1 and 6.
Explain how the environment can influence the development of inherited characteristics.	For You To Do BioTalk	Biology to Go Question 7.

Activity Overview

This activity introduces students to how characteristics are passed from parents to their offspring. They will observe how the environment can influence the expression of a particular trait. This activity will show that the effect of the environment is only temporary and that the normal expression of the trait will revert back when conditions are normal.

Preparation and Materials Needed

Preparation

Plan to start the planting component of the activity ahead of time. The seeds need time to germinate for at least two weeks. They will need to be watered during the germination period. You may do the planting component of the activity before you start **Activity 1**. Arrange a schedule for watering and observation. You may also start with **Activity 3** and **4** concurrently with the observation component of **Activity 2**.

You may wish to duplicate **Blackline Master Evolution 2.1: Kinds of Leaves for Germinating Tobacco Leaves.**

Materials/Equipment Needed

- blotting paper (substitutes could be filter paper or paper towel)
- 4 Petri dishes (per group)
- tobacco seeds
- black plastic bags or lightproof containers

Learning Strategies for Students with Limited English Proficiency

1. Point out new vocabulary in context. Practice using the words as much as possible.

   germinate seedlings factors albino
   lightproof

2. You may assign students to write the goals of the activity in their own words.

3. It is a good practice to read the procedure before the activity. You may also assign the students to summarize the procedure before the activity.

A Highway Through the Past

Activity 2 Is It Heredity or the Environment?

GOALS

In this activity you will:

- Observe how an inherited trait can be influenced by the environment.
- Distinguish between a genotype and a phenotype.
- Explain how the environment can influence the development of an inherited characteristic.

What Do You Think?

"You've got your mother's hair and your father's eyes." Almost everyone has heard about heredity at some point.

- **How are personal characteristics passed on from one generation to the next?**
- **Can a personal characteristic be changed?**

Write your answer to these questions in your *Active Biology* log. Be prepared to discuss your ideas with your small group and other members of your class.

For You To Do

In this activity you will use tobacco seeds from parents that carried the characteristics for albinism (no chlorophyll) but did not show it. Your observations will help you to understand that traits are inherited but are also influenced by the environment.

What Do You Think?

- Personal characteristics are passed from one generation to the next through the genes that are contained in the parents' chromosomes. Every individual has two sets of chromosomes or genes. Each parent contributes one set of chromosomes to the offspring.

- Personal characteristics can be temporarily influenced by the environment. This will go back to its natural form, when conditions revert back to normal. With plastic surgery, however, some personal characteristics can be cosmetically altered. This characteristic is artificial and will not be passed on to the offspring.

Student Conceptions

Some students will have the misconception that plants need sunlight to germinate. You may remind students that when seeds are planted, they are normally covered with soil. This clearly indicates that sunlight is not needed in the germination process.

Many students may think that the environment influences the traits that will be permanent and will be passed on from one generation to the next. This activity will clarify that misconception.

For You To Do

Teaching Suggestions and Sample Answers

Teaching Tip

In the absence of a lightproof container, cover the container with black plastic bags. A few minutes of exposure to light during watering will not alter the result significantly.

Teaching Tip

Ask the students to formulate a hypothesis on which containers will have the most number of albinos before they make their observation. Require them to write a conclusion on the activity based on their hypothesis.

1. Place blotting paper in the bottom of each of four Petri dishes. Moisten the paper, but be sure that it is not floating in water. Sprinkle about 40 tobacco seeds evenly over the surface of the paper. Keep the seeds a few seed lengths apart from each other.
2. Replace the covers of the dishes and place the dishes in a well-lighted place, but not in the direct sunlight. The temperature should be approximately 22°C.
3. Cover two dishes with a lightproof container.
4. Leave the other two dishes exposed to the light.
5. Let the seeds germinate for about a week, adding a few drops of water to the paper every other day or whenever the paper begins to dry.
6. On the tenth day, begin to make entries in your table of results.
   a) Make up tables on which to record your results. You may wish to use tables similar to the ones shown on the next page.
   b) Every day, record how many and what kind of seedlings you observe.
7. When all or most of the seeds have germinated in the darkened dishes (probably the twelfth day) remove the covering. Place these dishes in the light next to the others.
   a) Continue to record the appearance of the seedlings through the thirteenth day.
8. Study all the data you have accumulated.
   a) Try to draw any conclusions that you can from your data.
9. Using your data for the seedlings that were kept in the light all the time, answer the following questions:
   a) How might you explain the differences you observed?
   b) Are these differences caused by heredity or environment?

Wash your hands after handling the seeds. If mold forms in the dish, have your teacher dispose of the affected seeds.

9. a) The absence of light affected the color of the seedlings left in the lightproof container.
   b) The difference is definitely environmental because most of the seedlings that are albino are in the lightproof container.

NOTES

Kinds of Leaves from Germinating Tobacco Leaves (Dishes Continuously Exposed to Light)			
	Albino	Green	Percentage of albino each day
10th day			
11th day			
12th day			
13th day			

Kinds of Leaves from Germinating Tobacco Leaves (Darkened Dishes)			
	Albino	Green	Percentage of albino each day
10th day			
11th day			

Kinds of Leaves from Germinating Tobacco Leaves (Covering Removed from Darkened Dishes)			
	Albino	Green	Percentage of albino each day
12th day			
13th day			

10. Consider the seedlings that were kept in darkened dishes.
    a) How do the percentages of albino and green seedlings compare with the percentages of albino and green seedlings that were continuously exposed to light?
    b) What is the environmental factor that is varied in this activity?
    c) Is this difference in percentages of seedlings kept in the light and seedlings kept in the dark due to inherited or environmental factors?
    d) What do you think is causing the differences in the appearance of the seedlings that were in darkened dishes?
11. Consider the seedlings that were first in darkened dishes and then exposed to light.
    a) How do the percentages of green and albino seedlings compare with the percentages of green and albino seedlings in the other situations?
    b) What happened to the appearance of many of the seedlings after the cover was removed?
    c) Does this support your answer to **Step 10 (d)**?
    d) What are the effects of light upon seedlings that carry a certain hereditary characteristic?

10. a) The percentage of albinos will be greatest in the darkened dishes.
    b) The presence of light.
    c) Environmental.
    d) The absence of light.

Teaching Tip

Review with the students how to compute for the percentage of albino.

$$\text{Percentage of albino} = \frac{\text{number of albinos}}{\text{number of seedlings}} \times 100\%$$

You may allow the use of a calculator for computation.

11. a) The percentage of albino in the previously darkened dish is decreasing over time.
    b) The seedlings from the darkened Petri dishes start to resemble the seedlings that where exposed to light.
    c) Yes.
    d) In this activity the coloration of the seedlings were affected by the presence or absence of light.

Bio Talk

The Importance of Heredity and Environment

Why do offspring resemble their parents? Genetics, a branch of biology, tries to answer these types of questions about inheritance. Geneticists have found that most aspects of life have a hereditary basis. Many traits can appear in more than one form. A **trait** is some aspect of an organism that can be described or measured. For example, human beings may have blond, red, brown, or black hair. They may have tongues that they can roll or not roll. (Try it! Can you roll your tongue? Can your parents?) They may have earlobes that are attached or free. The passing of traits from parents to offspring is called **heredity**.

In *most* organisms, including humans, genetic information is transmitted from one generation to the next by deoxyribonucleic acid (DNA). DNA makes up the **genes** that transmit hereditary traits. Each gene in the body is a DNA section with a full set of instructions. These instructions guide the formation of a particular protein. The different proteins made by the genes direct a body's function and structure throughout life. **Chromosomes** carry the genes. They provide the genetic link between generations. The number of chromosomes in a cell is characteristic of the species. Some have very few, whereas others may have more than a hundred. You inherit half of your chromosomes from your mother and the other half from your father. Therefore, your traits are a result of the interactions of the genes of both your parents.

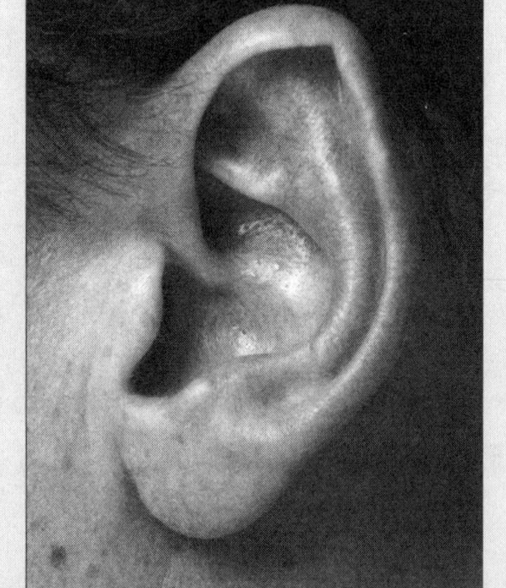

Bio Words

trait: an aspect of an organism that can be described or measured

heredity: the passing of traits from parent to offspring

gene: a unit of instruction located on a chromosome that produces or influences a specific trait in the offspring

chromosome: threads of genetic material found in the nucleus of cells

BioTalk

Teaching Tip

You may wish to further explain the difference between genotype and phenotype by showing how the genotype might look like for certain phenotypes. Assign capital letters to symbolize dominant genes and lower case to symbolize recessive genes. For example, for black coat color on mice let **B** equal the black coat factor and **b** the recessive white coat factor. A mouse with a black coat can either be pure black (**BB**) or heterozygous (**Bb**). An individual with a white coat will be pure white (bb). **BB**, **Bb**, and **bb** are the genotype representations, and black coat and white coat are the phenotype expressions.

Ask students to list five of their phenotypes. Ask them to also list all members of their family that share the same characteristics. Based on the information they gathered, the students can try to figure out their genotype.

NOTE: Not all traits are represented by just one. Just a hypothetical assumption that each trait in their example is represented by one gene.

A Highway Through the Past

Bio Words

dominant: used to describe the gene that determines the expression of a genetic trait; the trait shows up

recessive: used to describe the gene that is overruled by a dominant gene; the trait is masked

genotype: the genes of an individual

phenotype: the observable traits of an organism that result because of the interaction of genes and the environment

Gregor Mendel was the first person to trace the characteristics of successive generations of a living organism. He was an Augustinian monk who taught natural science to high school students. His origins were humble. However, his work was so brilliant that it took many years for the rest of the scientific community to catch up to it.

The modern science of genetics started with the work of Gregor Mendel. He found that certain factors in a plant cell determined the traits a plant would have. Thirty years after his discovery these factors were given the name genes. Of the traits that Mendel studied, he found that one factor, or gene, could mask the effect of another. This is the principle of dominance. He called the factor that showed up in the offspring **dominant**, and the factor that was masked **recessive**.

Genotype refers to the genes that an organism contains for a particular trait. The **phenotype** is the observable traits of an individual. Phenotype is a product of the interaction between the genotype and the environment.

All genes interact with the environment. Sometimes it is difficult to tell how much of a phenotype is determined by heredity and how much is influenced by the environment. A familiar example of how the environment affects the phenotype is the coloring of Siamese cats. The cats have a genotype for dark fur. However, the special proteins (enzymes) that produce the dark color work best at low temperatures. That is why Siamese cats have dark markings on their ears, nose, paws, and tail. These are all areas that have a low body temperature. Suppose a Siamese cat's tail were shaved and then kept at a higher than normal temperature. It would soon be covered with light-colored fur. However, these changes are temporary and only

NOTES

present if the environmental conditions are met. There are other examples of the influence of the environment on a phenotype. For a fair-skinned person, exposure to sunlight may produce hair that has lightened and a face full of freckles. Primrose plants are red if they are raised at room temperature, but white if they are raised at temperatures about 30°C. Himalayan rabbits are black when raised at low temperatures and white when raised at high temperatures.

The Siamese is a cat in which color is restricted to the points, i.e., nose, ears, legs, and tail. This is known as the Himalayan pattern. This coloration is a result of both hereditary and environmental factors.

Reflecting on the Activity and the Challenge

In this activity you saw that both heredity and environment contribute to the expression of a trait in a plant. You then read about how this also applies to animals. When thinking about how organisms are able to adapt, you must consider both inherited characteristics as well as the influence that the environment has on the organism. You will need to understand this for your **Chapter Challenge**. Also, in the next activity, you will investigate natural selection. Heredity and the environment both play a role in natural selection.

NOTES

Biology to Go

1. Distinguish between a genotype and a phenotype.
2. What is the difference between a dominant and a recessive gene?
3. A dominant gene for a specific trait is inherited along with a non-dominant gene for the same trait. Which gene's "building instructions" will be used to assemble the specific protein?
4. In guinea pigs black coat is dominant to white. Is it possible for a black guinea pig to give birth to a white guinea pig? Explain your answer.
5. Explain how both heredity and environment contribute to the expression of a trait in plants.
6. Review your observations from the activity. Comment on the following statement: heredity can determine what an organism *may* become, not what it *will* become.
7. Can the environment change the development of an inherited characteristic? Use your observations from this activity to justify your answer.

Inquiring Further

1. Analyzing a genetic condition

What is a genetic condition? Choose a condition from one of the more well-known conditions, such as achondroplasia, cystic fibrosis, hemophilia, Huntington's chorea, Marfan syndrome, dwarfism, Down syndrome, Fragile-X syndrome, Tay-Sachs disease, sickle cell anemia, neurofibromatosis, etc. You may wish to investigate a condition with which you are personally familiar. Construct a hypothetical family tree to do a pedigree analysis of the condition. (A pedigree is used to trace inheritance of a trait over several generations.)

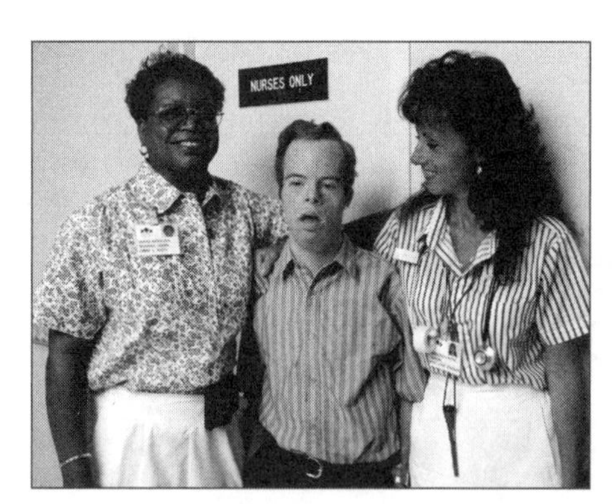

Down syndrome is caused by abnormal cell division in the egg, sperm, or fertilized egg. This results in an extra or irregular chromosome in some or all of the body's cells.

Biology to Go

1. Genotypes are the genes of individuals, and phenotypes are the individual's observable traits as a result of the interaction of the genes with the environment.

2. The factor that shows up in an offspring whenever it is present is the dominant gene and the factor that is usually masked is the recessive gene.

3. The dominant gene will always show an individual's characteristics and therefore will be used in the assembly of that specific protein.

4. Yes. If both parent genes have both the dominant and recessive gene for black coat, it is possible that the offspring will inherit only the recessive gene from each parent, resulting in a white coat.

5. The traits in plants are inherited from their parents. The gene is expressed in a particular way under normal conditions that the plant is accustomed to. However, changes in the environment may cause changes on how this gene is expressed.

6. The expression of a certain trait is predetermined by conditions the organism is accustomed to. The change in the environment may alter how that gene is expressed.

7. The environment can only temporarily change the expression of an inherited trait and will return to its normal expression when the conditions return to normal.

Inquiring Further

Analyzing a genetic condition

Students will easily find information about genetic conditions. You may need to show an example on how to construct a pedigree and how to draw information from it.

Blackline Master Evolution 2.1: Kinds of Leaves for Germinating Tobacco Leaves

Kinds of Leaves from Germinating Tobacco Leaves (Dishes Continuously Exposed to Light)			
	Albino	Green	Percentage of albino each day
10th day			
11th day			
12th day			
13th day			

Kinds of Leaves from Germinating Tobacco Leaves (Darkened Dishes)			
	Albino	Green	Percentage of albino each day
10th day			
11th day			

Kinds of Leaves from Germinating Tobacco Leaves (Covering Removed from Darkened Dishes)			
	Albino	Green	Percentage of albino each day
12th day			
13th day			

NOTES

ACTIVITY 3– NATURAL SELECTION

Background Information

Natural Selection

The theory of natural selection is generally attributed to Charles Darwin. He presented a lot of evidence from his travels to South America, the Galapagos Islands in particular. He observed differences and similarities among the finches from different islands in the Galapagos. He observed that only finches with certain characteristics can be found on a particular island. Finches with similar traits can be found living in adjacent islands. He also observed that the beaks of finches may have something to do with the availability of seeds. Finches with large beaks can be found where hard seeds are abundant, and finches with thin beaks can be found where small seeds are present.

Darwin was interested in how species change over time and how different species came about. He proposed that natural selection is the mechanism involved in the process.

In the process of natural selection, individuals with favorable characteristics are able to adapt to their particular environment. These individuals are able to reproduce and transfer their genetic characteristics to the succeeding generations. The individuals with less favorable characteristics will likely decrease in numbers and eventually become extinct. This explains why only certain type of finches can be found on certain islands in the Galapagos.

As the variations become isolated from each other and interbreeding is diminished, each group become specialized over time and become a new species.

In this activity, the students will simulate how camouflage, an adaptation they have learned in **Activity 1**, will help determine how a certain characteristic of a species living in a particular environment will disappear in succeeding generations. This activity will help students see how unsuitable characteristics disappear. This will help students formulate the idea of natural selection.

Goals and Assessments

Goal	Location in Activity	Assessment Opportunity
Investigate the process of natural selection.	For You To Do Steps 1 – 7	**For You To Do Part A:** Students are able to simulate natural selection **For You To Do Part B:** Students are able to speculate what happened to the moth population over time.
Describe the major factors causing evolutionary change.	BioTalk	**Biology to Go** Question 3.
Distinguish between the accommodation of an individual to its environment and gradual adaptation of a species.	BioTalk	**Biology to Go** Question 1.
Read about the meaning of theory in science.	BioTalk	

Activity Overview

In this activity students simulate predation in a hypothetical environment. The students record hypothetical survivors over four generations. From the data, students infer that favorable characteristics can increase an organism's chances for survival. They can infer that organisms with favorable characteristics are able to reproduce and multiply.

The students will also study the effect of building of a four-lane divided highway on the variations of a moth population over time.

Preparation and Materials Needed

Preparation

You may have to cut squares (about 1 inch square) of red (preferably bright red), white and newsprints. You will need 50 of each type per group of four students. You also need to prepare a class chart on the board to tally the class results.

You may wish to duplicate **Blackline Master Evolution 3.1: Predator and Prey** for the students to use to record their observations.

Materials/Equipment Needed (per group)

- a full sheet of newspapers
- 50 half-inch squares cutout of newspapers
- 50 half-inch squares cutout of white papers
- 50 half-inch squares cutout of red papers
- If the half-inch papers were not prepared for the students ahead of time, the following should be provided for each group:
- a pair of scissors
- ruler
- white papers
- newspapers
- red papers

Learning Strategies for Students with Limited English Proficiency

1. Point out new vocabulary in context. Practice using the words as much as possible.

keeper	suited	speckled
predation	unsuited	habitat

2. Ask the students to read through the procedure before they start the activity. You may want to require them to outline or summarize the procedure before the start of the activity.

3. English learners may have difficulty conceptualizing "the real world." You may want to clarify this for them.

Activity 3 Natural Selection

GOALS

In this activity you will:

- Investigate the process of natural selection.
- Describe the major factors causing evolutionary change.
- Distinguish between the accommodation of an individual to its environment and gradual adaptation of a species.
- Read about the meaning of a theory in science.

What Do You Think?

One hundred rabbits were trapped and introduced to an island with a huge diversity of plants. The rabbits had several noticeable variations. Thirty years later scientists returned to the island. They were amazed that although the number of rabbits was still around 100, the later generations did not vary as much as the earlier rabbits had.

- **What happened to the variations that were evident in the original species?**
- **How would you explain why the variations seemed to have disappeared?**

Write your answer to these questions in your *Active Biology* log. Be prepared to discuss your ideas with your small group and other members of your class.

For You To Do

In this part of the activity you will study the process of natural selection. You will work with a hypothetical population of organisms in a hypothetical environment.

What Do You Think?

- The rabbits with characteristics that were not favorable for survival in the new environment were not able to reproduce and transfer their characteristics to the succeeding generations.

- The variations disappeared because only the rabbits with the characteristics that best suit the island were able to reproduce and pass their traits.

Student Conceptions

Students may not be able to distinguish that results from a small sample will vary more than the results from a large sample. They would think that the data they get from an experiment is sufficient to make a generalization.

For You To Do
Teaching Suggestions and Sample Answers

Remind the students that they should always read through the steps of an activity before they begin.

Review what a hypothesis is before the start of the activity. Discuss or point out that a larger sample size will give a more accurate picture than a small sample size. This will make students understand why a class result is necessary in this activity.

The blank data table is available as **Blackline Master Evolution 3.1: Predator and Prey.**

A Highway Through the Past

You will use a sheet of newspaper as the environment. You will use paper squares to represent individual prey. You will be given a chance to capture five prey individuals. The remaining prey will reproduce. You will then have another chance to capture the prey.

Part A: Predator and Prey

1. Work in groups of four. One student (the keeper) sets up the environment before each round (generation). The other three in the group act as "predators." They remove prey from the environment.
2. Lay a sheet of newspaper flat on a table or floor.
3. Take at least 50 each of newspaper, white, and red paper squares (150 squares). Keep the three types separate, as each represents a different type of the *same* prey species. Some are brightly colored. The others are not. An example of such different populations is the species *Canis familiaris*, the common dog. Although dogs come in many different colors and sizes, they all belong to the same species. The paper squares represent individuals of different colors, but of the same species.
4. The keeper collects 10 squares from each of the three prey populations. The keeper then mixes them and scatters them on the environment while the predators are not looking. Each predator may look at the environment *only* when it is her or his turn. When it is not your turn, simply close your eyes or turn your back until the keeper indicates that it is your turn. When it is your turn, remove five prey individuals as quickly as you can. Continue in order until each predator has removed five prey individuals.
5. Shake off the individuals left on the environment and count these survivors according to their type. They represent generation 1.
   a) Enter the data for your group in a table similar to the one shown.
   b) Place the data on the chalkboard also, so a class total can be reached.

Generation		Paper-Prey Species		
		Newspaper Individuals	White-Paper Individuals	Red-Paper Individuals
1	Team			
	Class			
2	Team			
	Class			
3	Team			
	Class			
4	Team			
	Class			

Part A: Predator and Prey

1. Organize groups of fours or use the same grouping from the previous activity. Assign monitors for each group to get the materials needed for the activity. Allow students to select the keeper and the three predators. Assign group numbers.

5.

Teaching Tip

Monitor that each group are placing 10 each of white, red, and newspaper squares on the big sheet of newspaper.

Teaching Tip

Remind students to enter their data in the chart where it says "Team" in Generation 1. Ask the monitors to enter their data on the space provided on the chalkboard. Compute the class total after all the groups have entered their data.

It is a good strategy to require students to have their own individual charts even if it is a group activity. This will enable every student to closely follow what is going on during the activity.

6. Analyze your data for the first generation. Record answers to the following questions in your *Active Biology* log:

   a) Does any population have more survivors than the others?

   b) Write a hypothesis that might explain this difference.

   c) Consider your hypothesis. If it is valid, what do you predict will happen to the number of newspaper individuals by the end of the fourth generation? to the red-paper individuals? to the white-paper individuals?

7. The survivors will be allowed to "reproduce" before the next round begins. For each survivor, the keeper adds one individual of that same type. The next generation will then include survivors and offspring. This should bring the total prey number back up to 30.

8. The keeper scatters these 30 individuals on the habitat. Repeat the predation and reproduction procedures for three more generations.

   a) Calculate the change in the number of all three populations after each round.

9. Look at your data and analyze your findings.

   a) Does it take you a longer or shorter period of time to find one prey individual as you proceed through the generations? Give an explanation for this.

   b) How does the appearance of the surviving individuals compare with the environment?

   c) Is your hypothesis and your prediction in question supported, or do they need to be revised?

   d) Were the red-paper individuals suited or unsuited for this environment? Explain.

   e) Would you say this species *as a whole* is better adapted to its environment after several generations of selection by the predators? Explain.

10. Now think of the "real" world.

    a) Is appearance the only characteristic that determines whether an individual plant or animal is suited to its environment? If so, explain. If not, give several other characteristics.

    b) In your own words, what is natural selection? What role does reproduction play in your definition?

11. Now you may test some of your own ideas about natural selection.

    - What would happen if there was a change in the environment, such as a change in color of the habitat?
    - What would be the result if one type of paper square "reproduced" at a faster rate than the others?

Part B: Hypothetical Model

1. Examine the story shown in the pictures on the next two pages. It is purely a hypothetical model and not an actual situation that occurred.

6. a) Answers will vary. Remind the students to compare the class result with their group result.
   b) Organisms that are camouflaged (from **Activity 1**) have a better chance of survival.
   c) There will be more newspaper individuals at the end of the fourth generation.

7.

Teaching Tip

Make sure that the students have doubled only the surviving individuals and that the total is back to 30.

9. a) It is increasingly more difficult to find a prey as there are less red and white individuals and the newspaper individuals cannot be easily identified as they disappear in the background.
   b) The surviving individuals look like the environment.
   c) Answer will vary depending on the original hypothesis.
   d) They are unsuited in the environment because they can easily be seen by predators.
   e) Yes, the species as a whole is adapted to the environment. Only the individuals with characteristics that are not camouflaged in the environment become less in number, and the camouflaged individuals were able to reproduce and multiply.

10. a) No, some plants develop a bitter taste or poisonous substances to discourage plant eaters. A rodent has a small body and developed legs that allow it to run fast and escape to a nearby burrow.
    b) Individuals with characteristics that are better suited for a certain environment are able to reproduce and pass their characteristics to the succeeding generations.

11. Individuals with colors similar to the new environment will flourish, and the population of individuals with colors similar to the previous environment will be diminished.

    The rate of survival of these individuals will also increase if the predator population remains constant.

2. Discuss the following questions in your small group. Then answer them in your *Active Biology* log.

   a) What change took place in the environment of the original moth population?

   b) What change was produced in the moth population as a result of this environmental change?

   c) Provide evidence that indicates that the change in the moth population is not simply an

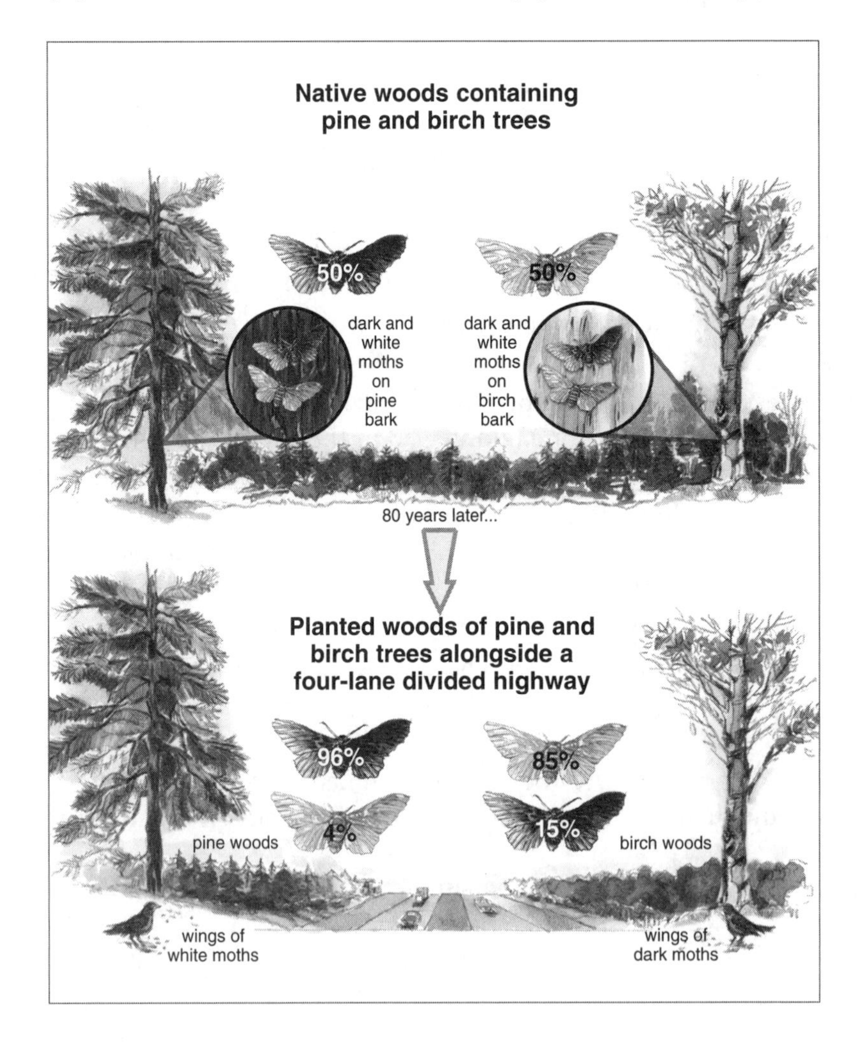

Part B: Hypothetical Model

2. a) The pine wood trees were separated from the birch trees by a four-lane highway.

   b) Decrease of white moth population on pine trees and increase of white moth population on birch trees. Increase of dark moth population on pine trees and decrease of white moth population on birch trees.

   c) The surviving dark moth can only interbreed with other dark moth, while the white moth can only interbreed with the white moth as movement between surviving moth populations in the birch trees and pine trees is restricted by the four-lane highway.

NOTES

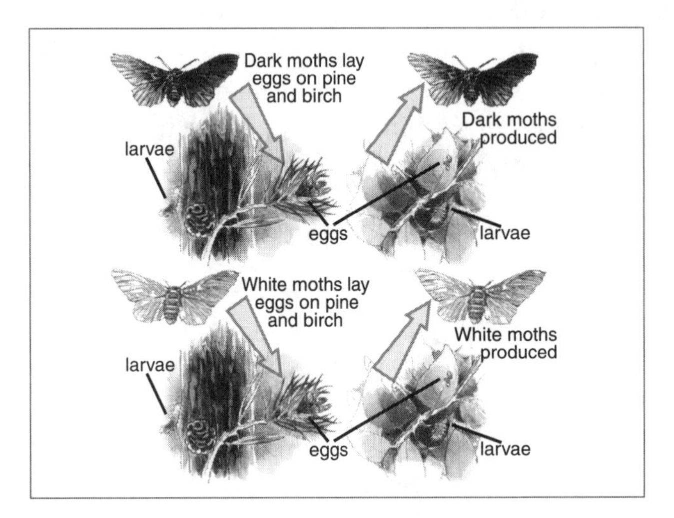

effect of the environment, but is really a hereditary population change.

d) Do you think that the change was a result of a change in reproductive capacity of the two kinds of moths? Do you think the change was a result of the survival of the moth best fit for the environment (selection pressure)?

e) What has happened to the frequency of the gene for speckled white color in the moth population now living in the pine woods?

f) What has happened to the frequency of the gene for speckled white color in the moth population living in the birch woods?

g) If environments change over a period of time, what must happen to populations if they are to survive?

h) If natural selection is responsible for the changes in frequency of black and of speckled-white moths in the two types of woods, what comparison can you make between the color of the favored type of moth and the color of the bark of the trees in each woods?

i) Assume there is benefit in protective coloration on the part of moths. What type of predators would you suspect to prey on moths?

j) What special abilities would these predators have to possess if they are really the agents of selection here?

k) Devise an experiment that would test this hypothesis.

d) Selection pressure is most likely the cause of the change. Movement of the moth population between the pine trees and birch trees are restricted by the four-lane highway. Since the white moth population is diminished on pine trees by predation, and the dark moth population is diminished on birch trees by predation, the surviving individuals can interbreed only with survivors with the same variations in each of the environments.

e) The frequency increased.

f) The frequency increased.

g) They should have the characteristics that best fit the new environment.

h) More dark moth survived in the dark pine trees, and more white moth survived in the white birch trees.

i) The predators cannot see individuals that are similar to the color of the environment.

j) They should be able to see organisms that are camouflage.

k) Put both moth populations in a background where both are visible.

Bio Talk

Theories in Science

The popular use and scientific use of the term "theory" are very different. Scientific theories attempt to provide explanations. Scientists make observations and then try to explain them. In popular terms you often hear the expression, "it's just a theory." That usually means that it is a guess. In scientific terms, a **theory** implies that an idea has been strongly supported by observations.

When scientists use the scientific method they often begin with questions from curious observations. They then develop hypotheses that can be tested experimentally. A **hypothesis** is a prediction between an independent (cause) variable and a dependent (result) variable. Hypotheses can either be supported or not, depending on the data collection. A hypothesis is not a guess. You developed and tested hypotheses in this activity. The hypothesis is then tested by further observations and experiments. Over time, if the observations and experiments satisfy the hypothesis, it becomes accepted as a scientific theory.

However, a theory is not the absolute truth. It only provides an explanation. The acceptance of a theory is often measured by its ability to enable scientists to make predictions or answer questions. A good theory provides an explanation that scientists can use to explain other observed events. Theories can be modified as new information becomes available or ideas change. Scientists continually "tinker" with a theory to make it more elegant and concise, or to make it more all encompassing.

Darwin's Hypothesis of Natural Selection

The theory of **evolution** owes much to the work of Charles Darwin. He presented his research in the mid-19th century. However, Darwin never labeled his hypotheses as "evolution." He was interested in how species change and how new species come about. His many years of work led to explanations that have proved to be valid. But Darwin was not the first to think that existing species might evolve into new ones. However, Darwin was a most believable scientist for two reasons. First, he amassed a great deal of evidence. He verified its accuracy and presented it in a convincing way. Second, his hypothesis stated *how* change in organisms might take place, a contribution no one else had made.

Bio Words

theory: a proven and generally accepted truth

hypothesis: an idea that can be supported or not supported by observations and experiments

evolution: a gradual change in the characteristics of a population of organisms during successive generations, as a result of natural selection acting on genetic variation

BioTalk

Teaching Tip

This may be a good opportunity to expand student knowledge on the difference of theory and a hypothesis. This is also an opportunity to clarify the difference between a hypothesis and a guess. After you discuss "cause and effect" in formulating hypothesis, then you may start requiring students to write their hypothesis in the "If... then" format.

From 1831 to 1836 Charles Darwin, a British naturalist, served aboard the H.M.S. Beagle on a science expedition around the world. The expedition visited places around the world, and Darwin studied plants and animals everywhere he went, collecting specimens for further study. In South America Darwin found fossils of extinct animals that were similar to modern species. On the Galapagos Islands in the Pacific Ocean he noticed many variations among plants and animals of the same general type as those in South America.

On November 24, 1859, the first edition of Darwin's *On the Origin of Species* was published. The book was so popular that its first printing was sold out in one day. There were, of course, many who disagreed with him. The theory of evolution has undergone many changes since Darwin's time. However, Darwin's original thinking still serves as a convenient introduction to the subject.

Here is his analysis:

- First, there are many differences among the individuals of every species. In a population, or group, of these individuals, variations occur. Usually it is safe to say that no two individuals are exactly alike. Darwin knew or suspected that many of the individual differences could be inherited.
- Second, the population size of all species tends to increase because of reproduction. One amoeba, for example, divides and produces two. These two divide, and the next generation numbers four. Then there will be 8, 16, 32, and so on.
- Third, this increase in the size of populations cannot go unchecked. If it did, the number of individuals of any species would outgrow the food supply and the available living space.

NOTES

- Fourth, it is obvious that this huge increase seldom occurs in nature. The number of organisms in a species does not continue to increase over long periods of time. In fact, the sizes of many populations seem to remain nearly the same over time. How can this be explained? Observations of natural populations show that many individuals die before they are able to reproduce.

Why do some individuals die early, but not others? Darwin thought there must be a sort of "struggle for survival." The individuals of a species "compete" for food, light, water, places to live, and other things important for their survival. The "struggle" or "competition" may be either active or passive. That is, sometimes animals actually fight for food or the opportunity to mate. In other cases, there is no direct fight or competition. The first animal that happens to find a suitable living area may settle there. This prevents the area from being used by others. In either case, individuals with certain characteristics, or traits, will survive and produce offspring more often than individuals without them.

While chasing prey, cheetahs often reach speeds of 70 miles per hour. Unfortunately, their great speed may not be enough for this species to survive. Scientists have found that wild cheetahs have virtually no genetic variation. Cheetahs suffer from inbreeding. This lowers their resistance to diseases and also causes infertility and high cub-death rates.

Consider, for example, how the African cheetah came to be such a fast runner. Cheetahs are hunters. They capture their food; mostly antelopes, gazelles, and birds, by first stalking near their prey. Then they run the prey down with a terrific burst of speed over a short distance. In any population of cheetahs, some can run faster than others. Those that run fastest are most successful in getting food. Those that are better at getting food also are more likely to survive.

NOTES

But survival is not the whole story. The characteristics that make an organism better able to survive in its environment are inherited. Therefore, those who survive are likely to pass on those characteristics to their offspring. For example, the surviving cheetah is likely to produce offspring with long, thin necks and powerful leg muscles, capable of great speed. Over many generations, then, one could expect an increase in the number of individuals that have these traits. The number with less beneficial characteristics would decrease. The organisms with the beneficial characteristics are likely to live longer and produce more offspring. Darwin called this process of survival and reproduction **natural selection**. Darwin thought that several factors were involved in natural selection:

1. The presence of variation among individuals in a population.
2. The hereditary basis of such variable characteristics.
3. The tendency of the size of populations to increase.
4. The "struggle for survival" (or competition for the needs of life).
5. A difference in the inherited characteristics that individuals pass on to succeeding generations.

Bio Words

natural selection: the differences in survival and reproduction among members of a population

Reflecting on the Activity and the Challenge

A change in the environment can have a large impact on the natural selection process. In this activity you investigated two situations. In the first, the "animals" that were best adapted to the environment were the ones to survive. In the second part you saw how a change in an environment could affect the natural selection process. Animals more suited to the changed environment would survive. You will need to explain the process of natural selection as part of your **Chapter Challenge**.

At one time some scientists believed that the necks of giraffes became long as a result of continually stretching to reach high foliage. Using what you know about natural selection, how would you explain the long necks?

NOTES

Biology to Go

1. What evidence supports the following idea: hereditary differences are important in determining whether or not an individual survives and leaves offspring?
2. What is the difference between natural selection and evolution?
3. What did Darwin emphasize as the major factors in causing evolutionary changes?
4. What did Darwin mean by natural selection?
5. Write a short paragraph expressing your ideas now of what happened to the rabbit population on the island in the **What Do You Think?** section.
6. Comment on the validity of the following statement: Breeders of domestic stock abandon natural selection. Only artificial selection plays an important role in animal-breeding programs.

Inquiring Further

1. Animal-breeding programs

What are the advantages and disadvantages to animal-breeding programs? Research and report on the pros and cons of human intervention in genetic processes.

2. Captive breeding

Captive breeding is one strategy used by governments and non-government organizations to preserve rare and endangered species. What are the advantages and disadvantages of captive breeding?

Biology to Go

1. The moth population, after the four-lane highway was built, is convincing evidence that genetic variations increase the chance for a species to survive. If there were only white moth, the moth population could be wiped out in the pine forest. If the environment drastically changed on both sides, after the other variation died out, all moth populations could be wiped out.

2. Natural selection is the environment favoring only individuals with characteristics that are able to adapt and reproduce to pass their characteristics to the succeeding generations. Evolution is the gradual change in the characteristics of a population of organisms in successive generations as a result of natural selection.

3. Darwin emphasized the following:
   - Presence of variation among individuals in a population
   - The hereditary basis of such variable characteristics
   - The tendency of the size of the population to increase
   - The "struggle for survival"
   - A difference in the inherited characteristics that individuals pass on to succeeding generations

4. Darwin meant that individuals with favorable characteristics to a certain environment will survive and produce offsprings more often than individuals without this characteristics.

5. The rabbits with characteristics that were not favorable for survival in the new environment were not able to transfer their characteristics to the succeeding generations. These characteristics were eventually lost and the variations were limited to the rabbits that were able to reproduce.

6. Breeders select only traits favorable to their purposes and raise domestic stocks in environmentally controlled environment. They abandon other variations of the trait which may limit the adaptive capability of the stock should there be changes in the environment.

Inquiring Further

1. Animal-Breeding Program

The domestic dog is one of the most commonly bred animal. Students can research on the advantages and disadvantages of different breeds of dogs and compare it with a dog of mixed breeds. You may also expand beyond animal breeding and explore plant-breeding programs. Recent issues on genetically altered food may be a good topic.

2. Captive Breeding

Several captive breeding programs are currently being undertaken and students can easily find information about them. The mountain gorilla, the panda, the bison and the California condor are examples of current breeding programs being undertaken. The students may also want to research the effect of salmon and fish hatcheries to repopulate rivers and creeks.

Blackline Master Evolution 3.1: Predator and Prey

Generation		Paper-Prey Species		
		Newspaper Individuals	White-Paper Individuals	Red-Paper Individuals
1	Team			
	Class			
2	Team			
	Class			
3	Team			
	Class			
4	Team			
	Class			

ACTIVITY 4– THE FOSSIL RECORD

Background Information

Fossils

Fossils are remains of ancient organisms found in rock layers. The remains of organisms may have been buried over time as layers of sand, soil, mud, and other materials are deposited and eventually hardened to form sedimentary rocks. During the process of sedimentation the soft materials of the organism's body were slowly replaced by the mineral deposits hardening to form the fossil. Some of these sedimentary rocks, where the fossil is embedded may have been lifted by geologic forces on Earth, eventually exposing the fossils through the process of erosion.

Some type of fossils may have been a result of an organism being entombed and preserved in ice during periods of glaciations, and are being revealed as they are exposed when the glaciers retreated. Others may have been stuck in resin, a gum-like material oozing from some tree trunks. The resin eventually hardened, forming what we now call amber, and preserved the organism inside. Complete samples of organisms are found preserved in resin or ice. The flesh that normally decays when an organism dies has resisted deterioration while inside the ice or the resin. Fossils found in rocks normally represent body parts that do not easily decay when an organism dies. Bones and exoskeleton of insects and other animals are some examples of these type of fossils.

Paleontologists discovered that remains of similar or related organisms are widespread within rocks layers of the same age all over the Earth. They also discovered that groups of organisms found in newer strata are more similar to modern organisms.

Various methods are used to date fossils. The most common method used for younger rock strata, is radiometric dating using the relative concentration of radioisotopes present in rocks. For rocks younger than 50,000 years, Carbon-14 is widely used. Other methods of dating rocks are also used: These include paleomagnetism, molecular clocks, continental drift, and changes in sea level.

The geologic history of the Earth is divided into eras. The eras are further subdivided into periods. Each period is defined by the prevalence of a certain group of organisms that existed during that particular time.

The changes in the type of fossils found in different layers of sedimentary rocks may have been due to physical changes on Earth that might have affected the environment of that time. These changes may be of extraterrestrial in origin like an asteroid strike on Earth. Environmental conditions on Earth may have been altered from time to time resulting in mass extinctions of organisms due to asteroid strikes. They could also be due to geologic changes on Earth over a long period of time, and also altering environmental conditions that can result in mass extinction. Continental drift, climatic changes, and volcanic eruptions may be some geologic events that have contributed to mass extinction on Earth.

In the activity, the students will simulate how fossils may have been formed. They are going to make models of fossils embedded in rock samples, ice, and resin.

Goals and Assessments

Goal	Location in Activity	Assessment Opportunity
Model ways in which fossils are formed.	**For You To Do** Part A **BioTalk**	
Explain the difference between body fossil and a trace fossil.	**For You To Do** Part B Human Population Steps 1 to 5	**Biology to Go** Question 3.
Describe the importance of fossils.	**BioTalk**	
Predict which animals are most likely to be found in the fossil record.	**BioTalk**	

Activity Overview

In this activity students model how fossils are made. They will make a model of how organisms can be preserved in rocks, resin, or ice. They will also model how trace fossils can be preserved in rocks.

Preparation and Materials Needed

Preparation

You will need to freeze several paper cups half-filled with water an hour before each class. The water should just be beginning to freeze when the students use them.

The stations should be set up before the students arrive for class.

Materials/Equipment Needed (per group)

For Every Student
- Goggles
- Laboratory gowns or aprons

Station 1
- plaster (a cupful per group)
- 1 clamshell (per group)
- petroleum jelly
- confetti
- 1 cup (per group)
- masking tape
- small hammer

Station 2
- 1 small paper plate (per group)
- hot-glue gun with glue sticks
- tweezers
- small seeds

Station 3
- modeling clay
- rolling pin or bottle
- plaster
- petroleum jelly

Learning Strategies for Students with Limited English Proficiency

1. Point out new vocabulary in context. Practice using the words as much as possible.

   clamshell
   confetti
   resin
   sedimentary rock
   abundance

2. It is always a good practice to have students write their own results, observations and interpretations of the activity even if the activity requires group effort and group discussion. Writing results and information in their own words help develop their writing skills.

Activity 4 The Fossil Record

GOALS

In this activity you will:

- Model ways in which fossils are formed.
- Explain the difference between a body fossil and a trace fossil.
- Describe the importance of fossils.
- Predict which animals are more likely to be found in the fossil record.

What Do You Think?

To hold a fossil in the palm of your hand is to have millions of years of history at your grasp. Fossils tell you about history, and like all good history, they help you to understand both the present and the past.

- **What is a fossil?**

Write your answer to this question in your *Active Biology* log. Be prepared to discuss your ideas with your small group and other members of your class.

For You To Do

In this activity you will have an opportunity to model different ways in which some fossils are formed. You will visit several stations.

What Do You Think?

Any evidence of past life preserved in sediments or rock.

Student Conceptions

Students are very familiar with fossils preserved in rocks. However, they may not have known about organisms preserved in ice or resin. Have a little background discussion before the activity about the cycle of cooling and global warming that formed vast amounts of glaciers over the world that could have preserved organisms in ice.

For You To Do

Teaching Suggestions and Sample Answers

Teaching Tip

It may be a good idea to start the activity on a Friday to allow time for the models in plaster or glue to dry over the weekend. Depending on the type of plaster, it may take more than a day for the plaster to harden.

Teaching Tip

The safety procedure should be reviewed before the start of the activity. Review cleanup procedure and responsibilities before the activity.

Teaching Tip

If time available for the activity is limited, assign the groups a day before the activity or use the same grouping of students from **Activity 3**. As in the previous activities, ask the groups to read through the procedure and divide the responsibilities among themselves.

Depending on the time available and the class size, you may want to organize how to conduct the activity in any of the following methods:

1. Prepare two sets of each station and rotate two sets of groups among the stations. Allow ten minutes per group per station for a total of forty minutes.
2. Instruct the group to divide the work. One student doing **Station 1**, and the other 3 students doing the other 3 stations.
3. Instruct the group to form subgroups of two, the first subgroup doing **Station 1** and **2** and the second subgroup doing **Stations 3** and **4**.

A Highway Through the Past

Station 1: Preservation in Rock

You will mold a clamshell in plaster to model how it might be preserved in rock.

1. Obtain a large paper cup. Identify the cup with the name of your group. With a paper towel, smear petroleum jelly over the inside of the cup.
2. Mix plaster in another container following the directions on the package. Work quickly to complete the next four steps.
3. Fill the cup half full of plaster.
4. With the paper towel smear some petroleum jelly on both surfaces of a clamshell. Gently press the clamshell into the plaster.
5. Sprinkle a few pieces of confetti over the surface, enough to cover about 50% of the surface.
6. Fill the rest of the container with plaster.
7. Let the plaster harden overnight.
8. In the next class, remove the hardened plaster from the container. Set the plaster on its side and cover it with a towel. With a hammer gently hit the plaster to break it at the layer of confetti.

Wear goggles when using the hammer. Be sure others nearby are also wearing goggles. To contain any bits of plaster, cover the fossil model with cloth or paper before hitting it.

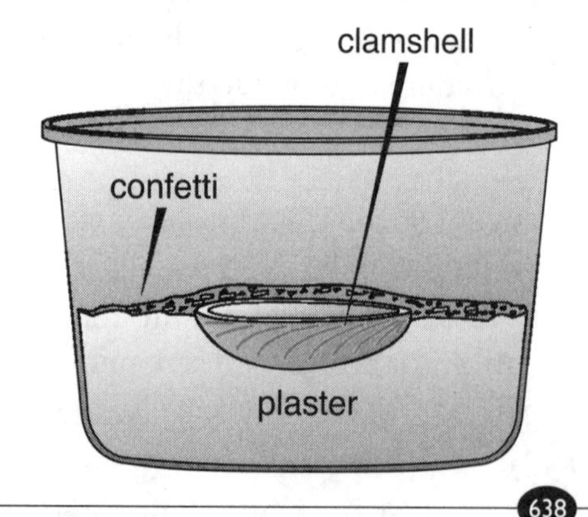

9. Observe your plaster molds and answer the following questions in your *Active Biology* log:
   a) What does the plaster represent?
   b) If you had never seen the clamshell, how would you figure out what the shell looked like by studying the fossil?
   c) Clamshells have two parts to their shell. How many possible imprints could a clamshell form?
   d) Why are fossils most often found in sedimentary rock formations?

Station 2: Preservation in Resin

You will encase a seed in glue to model how it might be preserved in a material like resin.

1. Obtain a small paper plate. Write your group's name on it. Use a paper towel to smear a small amount of petroleum jelly on a spot on the plate.
2. Using a hot-glue gun, put a bead of glue on the greased area of the plate.
3. Using tweezers, place the seed on the bead of glue. Add a few more drops of glue on top of the seed.
4. Let the glue harden overnight.

Wear goggles and be very careful when handling the hot-glue gun. Keep the hot part of the glue gun away from skin and flammable materials. Keep the glue away from skin, cloth, or other materials that may be damaged by it. Work on a surface that will not be damaged by the heat or the glue. Tell the teacher immediately of any accidents, including burns.

Station 1: Preservation in Rock

9. a) The plaster represents the rock where fossils are found.
   b) The imprint of the fossil represents the reverse form of the clamshell.
   c) Four, each shell half will form an outer and an inner imprint.
   d) Remains of organisms are covered by soil, sand, and mud over time. This hardens and eventually forms sedimentary rocks.

Teaching Tip

Show students several samples of actual fossils that they can compare with the models of fossils they produced. Point out similarities and differences between the real fossil and the model.

5. In the next class, remove the bead of glue and observe. Answer the following questions in your *Active Biology* log:

   a) Compare your preserved seed with a sample of amber provided by your teacher. How are they different? How are they similar?

   b) Explain how a seed might end up being preserved in the resin.

   c) Would you ever expect to see a large animal preserved in resin? Explain your answer.

   d) Which type of fossil would be easier to identify: one preserved in rock, or one preserved in resin? Explain your answer.

Station 3: Preservation in Ice

You will freeze a small object in a cup of water to model how organisms can be preserved in ice.

1. Your teacher will provide you with a paper cup half full of water that is beginning to freeze. Put your group's name on the cup.

Station 2: Preservation in Resin

5. a) The hardened glue is not as hard as the resin. The objects inside are visible in both.

   b) As the fruit opens and drops its seeds to the ground, the seed may have been stuck in the resin oozing from tree trunks. The resin will eventually harden.

   c) No. Trees do not form vast amounts of resin to be able to cover a large organism.

   d) The type of fossil preserved in resin. The actual body parts of the organism are preserved in resin while the fossil preserved in rocks may have been replaced by the minerals in the rock during the long period of sedimentation.

Teaching Tip

Ask students to collect resins from tree trunks in the school or their backyards. Pine trees are good candidates for resin deposits.

In the absence of a glue gun, use school glue. Ask the students to put the glue and the object in little containers instead of the small plate as the glue will flow. Please note that the school glue takes longer to dry.

A Highway Through the Past

2. Gently push the object under the surface of the ice.
3. Add more water on top of the object.
4. Let the water freeze overnight.
5. In the next class, remove the ice from the paper cup. Answer the following questions in your *Active Biology* log:

   a) How do you think an organism could end up being preserved in ice?

   b) What type of organisms could be preserved in ice?

Do not attempt to remove the object from the ice.

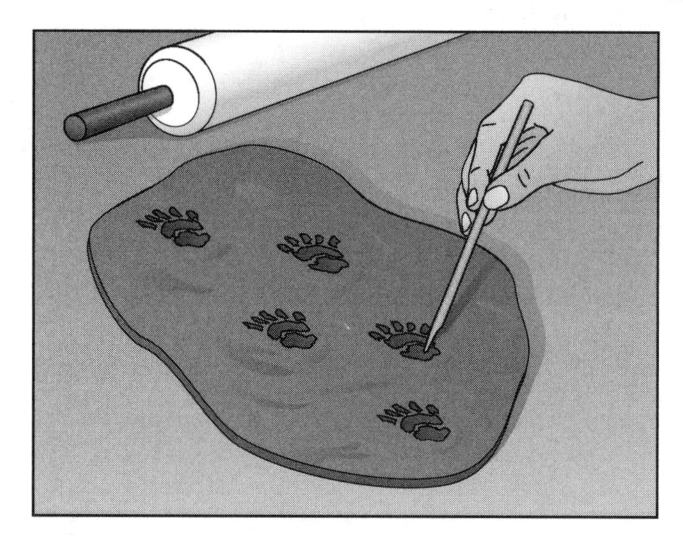

Station 4: Preserving Animal Traces

1. Flatten or roll out a piece of modeling clay to create a flat surface.
2. On the surface of the modeling clay, produce the pathway that an organism might leave in a muddy surface. Use your imagination to produce the pathway. You could represent anything from a worm crawling to a dinosaur trudging.
3. With a paper towel spread a small amount of petroleum jelly over the imprints you left in the modeling clay.
4. Mix plaster in a small plastic bag following the directions on the package. Cut the corner off the plastic bag. Squeeze enough plaster over the impression to fill the area.
5. Let the plaster dry overnight.
6. In the next class, remove the modeling clay from the plaster. Answer the following questions in your *Active Biology* log:

   a) In what kind of ancient environment(s) might you expect to have footprints formed?

   b) Once a set of fresh footprints have been made in the mud, what would have to happen to preserve them as rock?

Station 3: Preservation in Ice

5. a) The organism could be buried in a snowfall and preserved during glacial formation.
   b) All organisms, both small and large living in areas of glacial formation can be preserved in ice.

Station 4: Preserving Animal Traces

6. a) Wet muddy areas.
   b) They would have to harden over time to form sedimentary rocks.

Teaching Tip

You can show samples of different sedimentary rocks. Let students take little samples of the sedimentary rock to rub between their fingers. Ask students what material it is similar to. Knowing the composition of the sedimentary rock, ask the students to speculate what modern organism can likely become a fossil in that material.

You may also want to model deposition of sediments over time to help students determine the location of older rocks in relation to other rocks.

THE NATURE OF THE FOSSIL RECORD

Making Models

Scientists often make models to help them understand how living things work. Models can be small-scale structures that simulate what is found in nature. For example, a scientist might reconstruct the climatic conditions of 65 million years ago to uncover what might have happened to the dinosaurs. Another type of model could be nonliving structures that work in a similar fashion. The human heart is often understood from the model of a pump. Recently, scientists have begun using computers to make mathematical models. Unlike the structural models, these models only exist as numbers. In this activity you modeled the formation of fossils.

The Importance of Fossils

What does the fossil record tell you? Among a number of things, it tells you that species are not unchangeable. The species you see around you today are not the ones that have always existed. Fossils provide direct evidence that organisms are continually evolving. However, it is important to note that evidence of evolution is very different from the theories of evolution, which you read about in the previous activity. Fossils tell you that life forms on Earth have changed. The theories attempt to explain how and why these changes took place.

Fossil Formation

Fossils are preserved evidence of ancient life. Some fossils are called **body fossils**. These are the preserved parts of plants and animals. Fossils may also be **trace fossils**. These fossils are traces of the activities of plants and animals, for example, tracks, trails, or scratch marks.

As you investigated in this activity, fossils form as a result of many processes. For example, most animals become fossilized by being buried in sediment. The sediments then accumulate and consolidate to form rock. Molds are fossils formed from the impressions in soft sediment of shells or leaves, for example, or from footprints or tracks. Casts are replicas formed when a hollow mold is subsequently filled with sediment—mud, sand, or minerals. Sometimes an insect might

Bio Words

fossil: any evidence of past life preserved in sediments or rocks

body fossil: a fossil that consists of the preserved body of an animal or plant or an imprint of the body

trace fossil: any evidence of the life activities of a plant or animal that lived in the geologic past (but not including the fossil organism itself)

BioTalk

Teaching Tip

Show students several samples of actual fossils. Students can better relate what they read and see on books with actual specimens.

Teaching Tip

The chart on student page 643 is available as **Blackline Master Evolution 4.1: Major Divisions of Geologic Time.**

Teaching Tip

The graph on student page 644 is available as **Blackline Master Evolution 4.2: Comparative Abundance of Some Animals During the Mesozoic and Cenozoic Era.**

Assessment Opportunity

You may provide the students with the following questions to assess their understanding of reading the chart on page 643. The chart on student page 643 is available as **Blackline Master Evolution 4.1: Major Divisions of Geologic Time.** You can copy the chart for the students to use in answering the following questions:

1. In what period are modern humans prevalent? (*Quaternary period*)
2. What types of organisms became extinct during the Cretaceous period? (*Ammonoids and dinosaurs*)
3. What is most likely the index fossil used to define the Mesozoic era? (*Dinosaurs*)
4. What type of organism may have defined the Cambrian Explosion? (*Invertebrates)*

A Highway Through the Past

become trapped in a sticky substance called resin, produced by some types of trees. The resin hardens to form amber. The insect fossil is preserved in amber, often perfectly. At other times natural mummies form when organisms are buried in areas like tar pits and peat bogs or dry environments like deserts or certain caves. Organisms buried in

Body fossil.

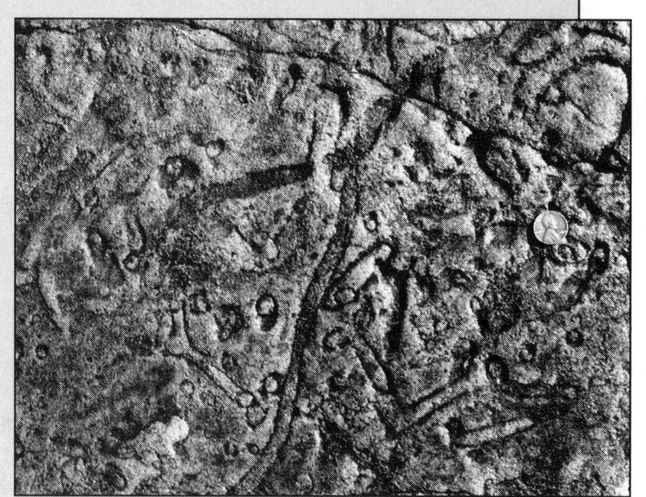

Trace fossil.

Cast.

NOTES

glacial ice also can remain preserved for thousands of years. Finally, the cells and pore spaces of wood and bone can be preserved if filled with mineral deposits, a process called petrifaction.

Not all organisms become fossils. To begin with, very few escape the food chain. They are either eaten by other organisms or are broken down by decomposers. Soft body parts decay very quickly. You know from experience that it takes little time for meat and vegetables to spoil if left out of the refrigerator. More resistant parts, such as the exoskeletons of insects, vertebrate bones, wood, pollen, and spores take much longer to decay. Thus, the likelihood of finding these in the fossil record is much greater.

Major Divisions of Geologic Time
(boundaries in millions of years before present)

Era	Period	Event	
Cenozoic	Quaternary	modern humans	1.8
	Tertiary	abundant mammals	65
Mesozoic	Cretaceous	flowering plants; dinosaur and ammonoid extinctions	145
	Jurassic	first birds and mammals; abundant dinosaurs	213
	Triassic	abundant coniferous trees	248
Paleozoic	Permian	extinction of trilobites and other marine animals	286
	Pennsylvanian	fern forests; abundant insects; first reptiles	325
	Mississippian	sharks; large primitive trees	360
	Devonian	amphibians and ammonoids	410
	Silurian	early plants and animals on land	440
	Ordovician	first fish	505
	Cambrian	abundant marine invertebrates; trilobites dominant	544
Proterozoic		primitive aquatic plants	2500
Archean		oldest fossils; bacteria and algae	

Time Not to Scale

The Fossil Record

Fossils typically form where sediments such as mud or sand accumulate and entomb organisms or their traces. The layers of hardened mud, sand, and other sedimentary materials are like a natural book of the Earth's history. Interpreting each layer is like reading the pages of a book. Unfortunately, there are many surfaces on the Earth where layers are not

NOTES

A Highway Through the Past

Bio Words

index fossil: a fossil of an organism that was widespread but lived for only a short interval of geological time

accumulating or where erosion is removing other layers. Thus, interpreting the layers is like reading a novel that is missing most of its pages. You can read the pages that are preserved and even group them into chapters, but much important information is missing from each chapter. Paleontologists (scientists who study fossils) use **index fossils.** These are fossils of organisms that were widespread but lived for only a short interval of geological time. They use index fossils to divide the fossil record into chapters. For example, dinosaurs are index fossils for the Mesozoic era, the unit of time that runs from roughly 245 million years ago (abbreviated Ma) to 65 Ma. In other

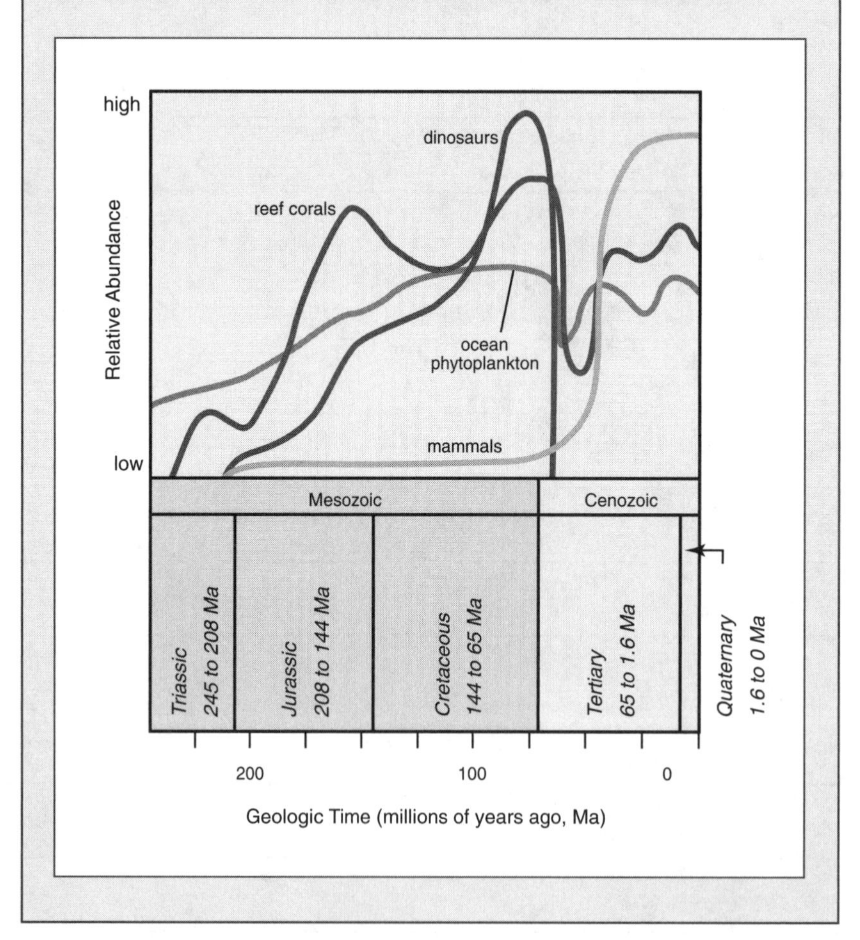

NOTES

words, all dinosaur species evolved and became extinct during the Mesozoic era. Whales, horses, and many other mammal groups, on the other hand, are index fossils of the Cenozoic era, the unit of time that runs roughly from 65 Ma to the present. Using this and additional fossil evidence, paleontologists infer that the Mesozoic and Cenozoic eras represent two of the major chapters in the history of life.

The graph on the previous page summarizes the distribution of four groups of animals during the Mesozoic and Cenozoic eras. (Please note that this graph is not drawn to scale along the vertical axis; for example, the peak in dinosaur diversity is comparatively a small fraction of present day mammal diversity.) Although the graph is a rough summary of just a small part of the fossil record, paleontologists can get a much more accurate picture of life's history by examining specific pages in the record. You will do this in the next activity.

Reflecting on the Activity and the Challenge

In this activity you modeled some of the ways in which fossils can form. You also read about different types of fossils and the incomplete nature of the fossil record. You may have now developed a sense of the importance of fossils. You can begin to appreciate what might be lost if fossil records were destroyed or disrupted. You will need to explain this at the town-hall meeting if you are representing one of the paleontologists.

1. What is the difference between a body fossil and a trace fossil?
2. What are the chances of an organism becoming a fossil? Explain your answer.
3. a) What is an index fossil?

   b) How do paleontologists use index fossils?
4. Use evidence from this activity to explain how the biosphere and the geosphere are connected.

Biology to Go

1. A body fossil is a type of fossil that consists of the preserved body of an organism or an imprint of the body of the organism. A trace fossil is any evidence of the life of a plant or animal that lived in the past, like a footprint preserved in mud that hardened over time.

2. Since very few organisms escape the food chain, the likelihood of an organism becoming a fossil is very low.

3. a) An index fossil is a fossil of an organism that was widespread but lives only a short interval of geologic time.
   b) They use index fossils to divide chapters of geologic time.

4. The clamshell gets imbedded in sand or silt where it lives (biosphere). The layer of sand and silt eventually hardens and forms sedimentary rock (geosphere) preserving the fossil of the clamshell.

Inquiring Further

1. Carbon-14 dating

How is it possible to determine the age of organic matter using carbon-14? Research to find the physical and chemical principles on which this technique is based. What are the limitations of carbon-14 dating?

Inquiring Further

1. **Carbon-14 Dating**
   Information on Carbon-14 dating is widespread. Students will have no problem researching and understanding the process of dating rock samples.

NOTES

Blackline Master Evolution 4.1: Major Divisions of Geologic Time.

Major Divisions of Geologic Time

(boundaries in millions of years before present)

Era	Period	Event	
Cenozoic	Quaternary	modern humans	1.8
	Tertiary	abundant mammals	65
Mesozoic	Cretaceous	flowering plants; dinosaur and ammonoid extinctions	145
	Jurassic	first birds and mammals; abundant dinosaurs	213
	Triassic	abundant coniferous trees	248
Paleozoic	Permian	extinction of trilobites and other marine animals	286
	Pennsylvanian	fern forests; abundant insects; first reptiles	325
	Mississippian	sharks; large primitive trees	360
	Devonian	amphibians and ammonoids	410
	Silurian	early plants and animals on land	440
	Ordovician	first fish	505
	Cambrian	abundant marine invertebrates; trilobites dominant	544
Proterozoic		primitive aquatic plants	2500
Archean		oldest fossils; bacteria and algae	

Blackline Master Evolution 4.2: Comparative Abundance of Some Animals During the Mesozoic and Cenozoic Era

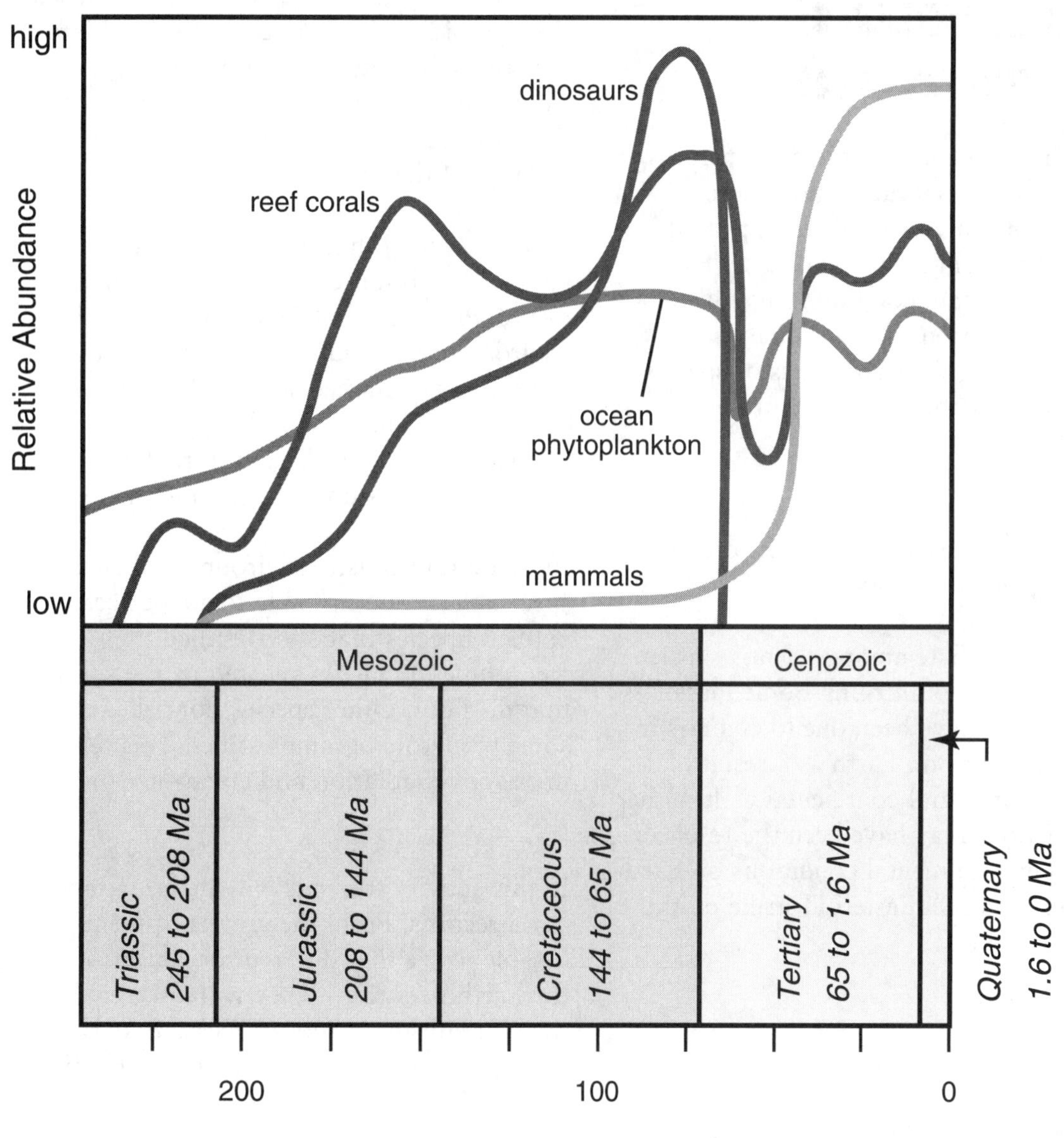

ACTIVITY 5–MASS EXTINCTION AND FOSSIL RECORDS

Background Information

The fossil records indicate there are periods when various new species flourished for a long period of time, and then disappeared completely in younger layers of rock. This indicates that certain conditions may have caused the extinction of these organisms. These periods of mass extinctions of species were followed by the emergence of new species.

These series of extinctions and emergence of new species were used to define the eras and periods of geologic history.

There is much debate among scientists as to the cause of mass extinctions. Some mass extinctions may have been due to changes in climatic conditions on Earth as a result of volcanic activities and continental drift. Other mass extinctions may have been the result of changes in environmental conditions on Earth in the aftermath of an asteroid strike on the surface of the Earth.

In the impact hypothesis, the primary evidence that points to a possible extraterrestrial in origin was a layer of iridium in the 65 million years rock samples around the globe that coincide with the time of an asteroid strike in the Yucatan Peninsula. Iridium does not naturally occur on Earth. This iridium may have been deposited when the asteroid was vaporized and spread into the atmosphere at the time it crashed. Advocates argue that this may have caused the Earth to darken for years, resulting in the collapse of the food chain.

Others contend that the mass extinction may have been a result of local geologic changes over a long period of time that altered the environmental conditions on Earth. Other researchers combined the two hypotheses. The asteroid impact may have hastened extinction that was already started by geologic changes on Earth during this period.

As a result of drastic environmental changes, species that were not able to adapt became extinct. Species that survive filled environmental niches vacated by the species that died out. Other species flourished as some predators became extinct. This resulted in adaptive radiation and emergence of new species.

In this activity students will analyze graphs of fossil records. From the graphs, students will be able to see that different organisms inhabit the Earth over time. They will also see that these organisms eventually cease to exist and are replaced by other forms of the organism.

Goals and Assessments

Goal	Location in Activity	Assessment Opportunity
Investigate fossil data for evidence of mass extinction and adaptive radiation.	For You To Do BioTalk	Biology to Go Question 1.
Explain the meaning of mass extinction.	BioTalk	Biology to Go Questions 2 and 3.
Describe the meaning of niche in an ecosystem.	BioTalk	Biology to Go Questions 2 and 4.

Activity Overview

Students analyze graphs of fossil data that represent the Cretaceous and Tertiary periods. From the graphs, the students will be able to see evidence of mass extinction and adaptive radiation.

Preparation and Materials Needed

Preparation

Prepare several different species of brachiopods for the students to analyze.

Materials/Equipment Needed

- Brachiopod fossils or shells

Learning Strategies for Students with Limited English Proficiency

1. Point out new vocabulary in context. Practice using the words as much as possible.

   paleontologist inference

2. Form collaborative groupings to conduct the activity. Allow time for group discussion and analysis. Assign discussion leaders to solicit contribution from all members of the group As in previous group activities, require individual reports.

Activity 5 Mass Extinction and Fossil Records

GOALS

In this activity you will:

- Investigate fossil data for evidence of mass extinction and adaptive radiation.
- Explain the meaning of mass extinction and adaptive radiation.
- Describe the meaning of niche in an ecosystem.

What Do You Think?

Sixty-five million years ago the curtain came down on the age of dinosaurs when a catastrophic event led to their mass extinction.

- **What type of disastrous event could have led to the extinction of such a large group of animals?**
- **Did any other life forms become extinct at this time in geological history?**

Write your answer to these questions in your *Active Biology* log. Be prepared to discuss your ideas with your small group and other members of your class.

For You To Do

In this activity, you will investigate fossil data from those "pages" that represent the boundary between the Cretaceous and Tertiary periods (about 65 million years ago).

1. Your teacher will divide the class into groups of three or four students. With the other members of your group,

What Do You Think?

- Examples of the possible cause of mass extinctions are long periods of volcanism and large asteroid falling into the Earth that changed the atmospheric conditions.
- Yes, many species of organisms other than the dinosaurs became extinct at this time. About half of marine life and many families of plants and animals were exterminated during this period.

Student Conceptions

Although it is clearly stated in the activity that **Graph A** represents only fossils from one location in Denmark, students may make the assumption that it represents brachiopod fossils worldwide. You may need to clarify this before the start of the activity. A better picture of what the data represents will help them figure out the re-emergence of some of the species.

examine the six brachiopod fossils. Brachiopods are a group of marine animals.

Wash your hands after handling the fossils.

a) What characteristics might paleontologists use to assign these fossils to different species?

b) What characteristics might paleontologists use to assign these fossils to one group?

2. Now examine **Graph A.** It plots the ranges of 50 different species of brachiopods across 15 m of sedimentary rock at one location in Denmark. This is one of the few places in the world that contains a continuous record of layers that represent the boundary between the Cretaceous and Tertiary periods. Using a technique known as *magnetostratigraphy*, geologists infer that each meter of this sedimentary sequence represents 0.1 million years of history. The point "0 m" represents the boundary between the Cretaceous and Tertiary systems, 65 Ma (millions of years ago). Paleontologists sampled fossils from the locations shown along the left axis.

a) Which species became extinct at the Cretaceous-Tertiary (K-T) boundary?

b) Which species evolved after the K-T boundary?

c) Which species appear to have become extinct and then reappeared later?

d) What conclusions can you draw from this graph?

e) What are the limitations of the data shown in **Graph A**? (Hint: recall the processes by which fossils are preserved.)

Graph A: Range of Different Species of Brachiopods

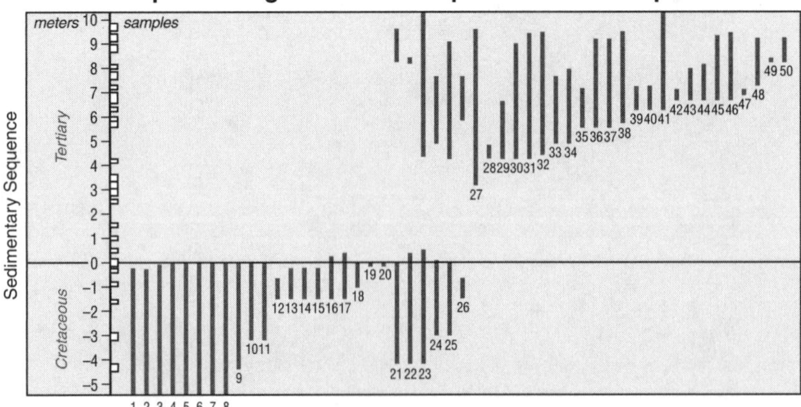

For You To Do
Teaching Suggestions and Sample Answers

1. a) Characteristics unique only to certain groups of individuals.
   b) Characteristics that are only common to this group of organisms.

2. a) Species 1 – 26.
   b) Species 27 – 50.
   c) Species 20 – 26.
   d) Something happened during this period that caused several species of brachiopods to become extinct.
   e) The data shows only fossils that are found. Some species may have survived but no fossil evidence was found during that period. Representatives of the species may have survived elsewhere, and moved into the area later. This is evident in species reappearing later in the Tertiary period.

Teaching Tip

Review with the students how to read information from different types of graphs. Students may get confused reading **Graph A.** You may need to go over how the information was presented on the graph before the start of the activity.

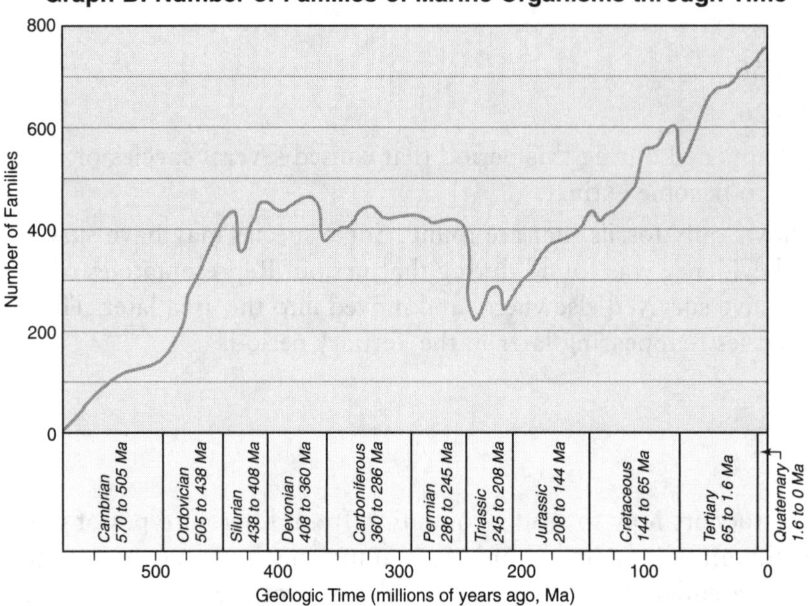

3. Paleontologists have compiled similar data on the ranges of existence of numerous other organisms during the Cretaceous and Tertiary periods. Examine **Graph B**, which shows a set of data assembled to illustrate the number of *families* (groups of closely related species) of marine animals through geological time.

   a) When was the number of families the greatest?

   b) Has the growth in the number of families been steady? Explain your answer.

   c) What do the dips in the graph represent?

   d) What inferences can you draw from this graph?

   e) What are the limitations of the graph shown in **Graph B**?

4. Now re-examine **Graph B**. Locate the times of the five greatest decreases in the number of families. Discuss why this might represent mass extinctions. Locate the times of the five greatest increases in the number of families. Discuss why this might represent adaptive radiations. (Adaptive radiation describes the rapid changes in a single or a few species to fill many empty functions in an ecosystem.)

   a) In your *Active Biology* log, construct a chart that summarizes your findings. Your chart should have two vertical columns, one labeled "Times of Mass Extinction" and the other labeled "Times of Adaptive Radiation." Fill in the chart with the estimated date that each event began and the name of the time period (e.g., "beginning of the Devonian period, roughly 410 Ma").

3. a) During the Quaternary period.
   b) From the Triassic period to the Quaternary period, there is a proportional increase in the number of families. The graph shows that the number of families remained steady from the Ordovician to the end of the Permian period.
   c) The dips in the graph show decrease or extinction of some families.
   d) The number of different families of marine organisms is increasing steadily over time punctuated by slight decreases over time.
   e) The graph shows only records of organisms with fossils.

4.

Time of Mass Extinction	Time of Adaptive Radiation
End of the Ordovician Period roughly 438 Ma	Beginning of the Cambrian Period roughly 570 Ma
End of the Devonian Period roughly 360 Ma	Beginning of the Silurian Period roughly 440 Ma
End of the Permian Period roughly 245 Ma	Beginning of the Triassic Period roughly 245 Ma
End of the Triassic Period roughly 208 Ma	Beginning of the Jurassic Period roughly 208 Ma
End of the Cretaceous Period roughly 65 Ma	Beginning of the Tertiary Period roughly 65 Ma

Teaching Tip

Provide rulers so students can easily identify the boundaries of the periods of adaptive radiation and mass extinction.

A Highway Through the Past

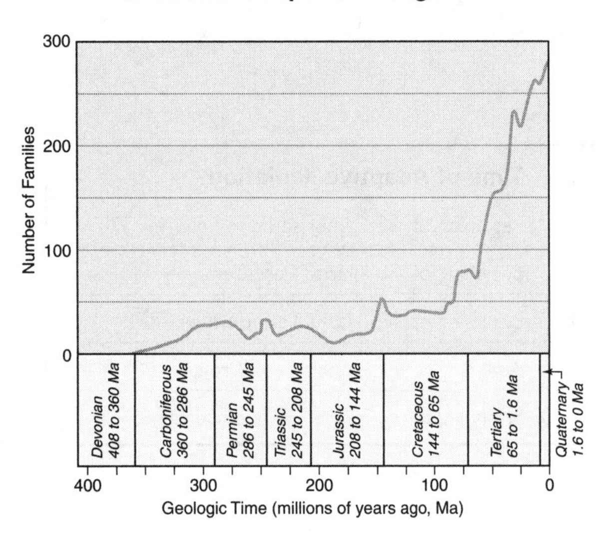

5. Now analyze **Graph C**, a graph constructed to show the number of families of terrestrial tetrapod families (land animals with four limbs) through geological time. Locate the greatest extinction events and adaptive radiations.
   a) Compare these events to the events listed in your chart for **Graph B.** Propose a hypothesis to account for the differences and similarities in these two graphs.
6. Consider the pattern of extinction and adaptive radiation in **Graph B** and **C.**
   a) How might adaptive radiation be related to mass extinctions? (Hint: consider how life on Earth might be different if dinosaurs still existed.)

BioTalk

Making Inferences in Science

Have you ever wondered how scientists know so much about dinosaurs? No human ever saw a dinosaur eat or run. The huge lizards disappeared from Earth about 65 million years ago. No fossil evidence of the human species, Homo sapiens, appears before 500,000 years ago.

The skeletons of dinosaurs have been reconstructed using fossil records. The skeletons provide indirect evidence of how the dinosaur might have lived. Evidence from the skull of a dinosaur may indicate that the dinosaur might have been a meat eater. The premise that this dinosaur killed other dinosaurs is called an inference. No one ever saw the dinosaur eating meat, the evidence to support this conclusion came from examining the skull shape and the structure of the teeth. Unlike a hypothesis, an inference cannot be tested.

5. **Graph B** and C show the same periods of mass extinction and adaptive radiation. Periods of adaptive radiation for terrestrial tetrapods are not as pronounced as the marine organisms up to the Cretaceous period. Adaptive radiation for tetrapods exploded during the Tertiary period. The overall increase in adaptive radiation during the Tertiary and Quaternary periods may be due to the lesser incidences of mass extinction.
6. The removal of some species during mass extinction can result in the extinction of other species on the food chain. This will also present opportunities for survivors to flourish and fill empty ecological niches that will eventually lead to adaptive radiation.

Teaching Tip

You may wish to illustrate what will happen on the food chain if some species are removed.

BioTalk

Teaching Tip

Students will have a hard time understanding what an ecological niche is. You can start by asking the students examples of animals and where they live. Where they live is their habitat. Continue by asking them what is the role of this animal in relation to other animals in the same habitat. Remind the students that the role or function of that animal, in relation to the other organisms and their environment, is its ecological niche.

Assessment Opportunity

1. What is the difference between extinction and mass extinction? (*Extinction is the disappearance of a species and mass extinction is the disappearance of a large number of species.*)
2. How are mass extinction and adaptive radiation related? (*Mass extinction can result in a lot of empty ecological niches. These empty ecological niches can be filled by survivors resulting in the emergence of new species or adaptive radiation.*)

Mass Extinction and Adaptive Radiation

Extinction is the total disappearance of a species. Extinction means that not a single organism of the species lives anywhere on Earth. The fossil record is a virtual graveyard of extinct species. It is strewn with the fossilized remains of millions of extinct species. David Raup, a paleontologist at the University of Chicago, notes that "only about one in a thousand species [that have lived on Earth] is still alive—a truly lousy survival record: 99.9 percent failure!"

Even more striking, however, is the fossil evidence of **mass extinctions**. These are episodes during which large numbers of species became extinct during short intervals of geological time. In geological time a few million years or less is a short period! The extinction of one species often has a domino effect. If one species vanishes, so do many others. Yet mass extinctions can present new opportunities to survivors.

Those best able to survive fill empty **niches**. (An ecological niche is the function a species plays in an ecosystem.) Plants and animals that have the greatest genetic variation are most often best able to fill these empty "spaces." This process is called **adaptive radiation**. In this activity, you investigated evidence of mass extinctions and adaptive radiations by analyzing data from the fossil record. Rapid evolution can also occur when a species moves into a new area. Natural variation within a species makes it easier for the species to adapt to different environments.

One remarkable mass extinction event occurred at the boundary between the Cretaceous and Tertiary periods, roughly 65 million years ago. This boundary separates the age of the reptiles and the age of the mammals. Geologists recognized this event over one hundred years ago when they realized that there was a striking change in the types of fossils deposited on either side of this boundary. This is where the language of science may become difficult to follow. However, no matter how it is said, the concepts are the same. This boundary

Bio Words

extinction: the permanent disappearance of a species from Earth

mass extinction: the extinction of a large number of species during short intervals of geological time

niche: the ecological function of a species; the set of resources it consumes and habitats it occupies in an ecosystem

adaptive radiation: the diversification by natural selection, over evolutionary time, of a species or group of species into several different species that are typically adapted to different ecological niches

NOTES

A Highway Through the Past

Model of a Brachiosaurus.

also separates two eras. These two eras are called the Mesozoic and Cenozoic. Dinosaurs were prevalent during the Mesozoic Era and extinct during the Cenozoic Era. The last segment of the Mesozoic Era is called the Cretaceous Period. The first segment of the Cenozoic Era is called the Tertiary Period. The abbreviation for the boundary between the Cretaceous and Tertiary periods is often referred to as the K-T boundary, where K is the abbreviation for the German form of the word Cretaceous. You may also hear of this time referred to the Mesozoic and Cenozoic boundary. No matter what you call it, there were dinosaurs before and there are no dinosaurs now, and it happened about 65 million years ago!

Moreover, at the end of the Cretaceous period virtually all plant and animal groups were lost from Earth, not just the dinosaurs. Yet, the beginning of the Tertiary period marks the start of the adaptive radiation of mammals.

The ultimate cause of the mass extinction at the Cretaceous/Tertiary boundary is still a debate among scientists. However, more and more evidence suggests that a meteorite impact caused the mass extinction. The impact of the meteorite created a chain of devastating environmental changes for living organisms.

Reflecting on the Activity and the Challenge

In this activity you had an opportunity to see that life forms that dominated the Earth many (geological) years ago are not here now. You also learned that the evidence to support this fact is found in fossil records. The new organisms that evolved to fill the ecological place of the extinct organisms are also part of fossil records. You probably have developed an even further appreciation of the importance of fossil records. You may wish to argue for a delay of construction of the new highway whether you represent the paleontologists or a concerned citizen.

NOTES

Biology to Go

1. Explain the meaning of adaptive radiation in your own words.
2. What evidence do scientists use to support the idea of mass extinctions?
3. After a mass extinction, which organisms are most likely to survive? Explain your answer.
4. Explain why the extinction of one species can have a domino effect on an ecosystem.

Inquiring Further

1. The Cretaceous and Tertiary boundary event

Research the proposed causes of the mass extinction during this time period. Provide at least two explanations. Which one do you think is more plausible?

Biology to Go

1. Adaptive radiation is the emergence of new species from a common ancestor by natural selection as they adapt to different environments.

2. The absence of fossil evidence of a large number of species over a period of geological time is an evidence scientists use to support mass extinction.

3. Organisms with a lot of genetic variation that are able to fill empty ecological niches are most likely organisms that survive after a mass extinction.

4. The extinction of one species affects the food chain. This will eventually lead to the extinction of other species that are dependent on them.

Inquiring Further

1. The Cretaceous and Tertiary boundary event.

Students can find various new articles on the subject matter. Recent Science magazine articles are reinforcing the impact hypothesis in the Yucatan Peninsula. This can be a good research project that the students can undertake. You may assign students to study different eras or periods in the geologic history with emphasis on the characteristics of organisms that are prevalent during the period. The research may be presented in class using oral presentation or a poster gallery.

A Highway Through the Past

Biology at Work

Pat Holroyd

Paleontologist, University of California Berkeley Museum of Paleontology

"Believe it or not," says Pat Holroyd, "the science of paleontology owes a lot to the oil industry and construction projects in general." This statement may be surprising to you, coming from one of the best paleontologists in the world.

Holroyd's position at the University of California Berkeley Museum of Paleontology (UCBMP) has her in the field looking for rare fossils and overseeing construction projects as much as 12 weeks a year. "Under California law every construction project must produce an environmental impact statement before ever breaking ground," she explains, "and that must include a paleontology component. So, if someone wants to put in a transmission line or a road they must first look at how all the natural resources might be affected."

Yet, according to Holroyd, that rarely means development projects are blocked or even seriously delayed. "Mitigation is the word we like to use," she says. "When excavation begins, a paleontologist will be there to see if anything is coming out of the hole. What usually happens is that the paleontologist will jump in the hole and excavate for as long as she or he has to and then the project continues." As Holroyd explains, "Most important fossil finds do not happen despite development, but rather because of them. Most of the fossils we've found are only found because of construction projects. "The fact is that there are only a few thousand paleontologists in the entire world and they don't have the money to dig a 30-meter hole in the ground."

Pat Holroyd examines a juvenile sea turtle skull, *Puppigerus camperi.*

As a child, she loved playing in the dirt, but thought more about discovering pyramids than fossils. "It wasn't until college that I became interested in paleontology," Holroyd says. After graduating from the University of Kansas, she received a Ph.D. from Duke University in biological anthropology and anatomy. "I usually just say my degree is in paleontology," Holroyd laughs. During graduate school she worked with the U.S. Geological Survey in Denver and continued there for a year after graduation before moving over to UCBMP, where she's been happily digging in the dirt ever since.

In the larger sense, Holroyd's work is the study of small mammals and the effects of global warming. She and other scientists are actively trying to determine what a phenomenon like that does to a whole ecosystem, in an attempt to see what might be happening now. "If there is global warming now, then it's important to look at a period in the past when the globe went through similar changes," she explains. "Almost everything that we, as scientists, look at in terms of the impact humans are having or might have on the environment are things that we can find examples of in the fossil record."

NOTES

Alternative End-of-Chapter Assessment

Multiple Choice

For numbers 1–5, choose the best answer.

1. A species is a group of organisms that:
   a) can interbreed but will produce sterile offspring
   b) can interbreed and produce fertile offspring
   c) occupy a given area
   d) have been introduced into an environment

2. Which of the following is an example of a trace fossil?
   a) a fossil of a microscopic organism
   b) an insect in a piece of amber
   c) a track left by a worm
   d) a small part of a bone

3. Characteristics of a species that makes its member better able to live and reproduce in their environment are known as:
   a) favorable adaptations
   b) dominant genes
   c) biotic factors
   d) recessive traits

4. According to Charles Darwin, which of the following factors are NOT involved in the process of natural selection?
   a) the presence of variation among individuals
   b) the hereditary basis of variable characteristics
   c) the struggle (competition) for survival
   d) the use and disuse of characteristics passed onto offspring

5. The diagram below shows undisturbed sedimentary rock at the bottom of an ocean.

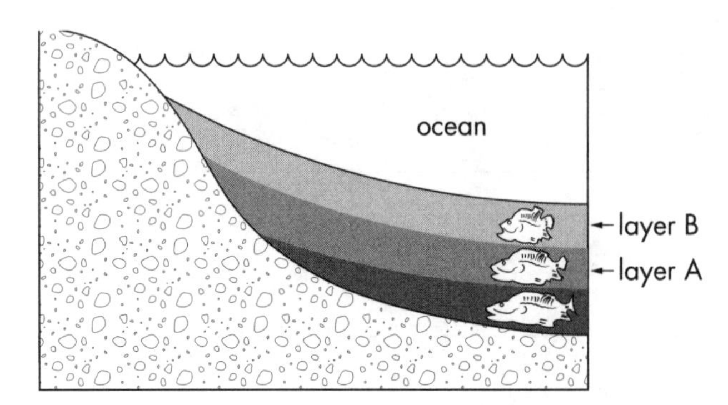

The fossils found in layer B resemble the fossils found in layer A. This similarity suggests that:

a) the fossils in layer B were formed before the fossils in layer A

b) modern forms of life may have evolved from earlier forms of life

c) vertebrate fossils are only found in sediments

d) the fossils in layer A must be more complex than those in layer B

True or False

For numbers 6–10, indicate if the statement is true (T) or false (F).

6. An animal that is radially symmetrical is symmetrical along a line running down the length of the animal.
7. Plants cannot display behavioral adaptations.
8. Camouflage is an adaptation that can help protect an organism from predators, but cannot help an organism capture prey.
9. A trait is an aspect of an organism that can be described or measured.
10. The time period following a mass extinction is followed by a rapid evolution of new species.

Written Response

11. Explain the difference between a structural and a behavioral adaptation.
12. How is body symmetry related to the speed at which an animal moves?
13. Explain how the study of fossils in sedimentary rock can shed light on evolutionary history.
14. The Russian or Himalayan rabbit is white with black extremities. After exposure to a low temperature for a period of time, a patch of fur on the back of the rabbit also turns black. Explain how this could happen.
15. Explain adaptive radiation.

Answers to Alternative End-of-Chapter Assessment

1. (b)
2. (c)
3. (a)
4. (d)
5. (b)
6. F
7. F
8. F
9. T
10. T

11. Structural adaptations are body structures that are advantageous to an organism's chances for survival. Behavioral adaptation is what an organism does to better its chances for survival.

12. Animals that display radial symmetry are not highly adapted for movement.

13. Sedimentary rock is formed when soil and rock particles are laid down in layers. Newer layers are formed on top of older ones. If the layers of sediment have not been disturbed more recent fossils are found near the surface. This allows scientists to compare the older fossils with more recent ones to see how a species has evolved.

14. This is an example of the effect of the environment on gene action.

15. Adaptive radiation is the diversification by natural selection, over evolutionary time, of a species or group of species that are typically adapted to different ecological niches.

NOTES